ECONOMIC ISSUES, PROBLEMS AND PERSPECTIVES

CARBON TAXES

ELEMENTS, CONSIDERATIONS AND OBJECTIVES

Economic Issues, Problems and Perspectives

Additional books in this series can be found on Nova's website under the Series tab.

Additional E-books in this series can be found on Nova's website under the E-book tab.

Energy Policies, Politics and Prices

Additional books in this series can be found on Nova's website under the Series tab.

Additional E-books in this series can be found on Nova's website under the E-book tab.

ECONOMIC ISSUES, PROBLEMS AND PERSPECTIVES

CARBON TAXES

ELEMENTS, CONSIDERATIONS AND OBJECTIVES

NEIL VIVEIROS
EDITOR

nova publishers
New York

For permission to use material from this book please contact us:
Telephone 631-231-7269; Fax 631-231-8175
Web Site: http://www.novapublishers.com

Library of Congress Cataloging-in-Publication Data

ISBN: 978-1-62618-148-9

Published by Nova Science Publishers, Inc. † New York

CONTENTS

PREFACE

The federal budget deficit has exceeded $1 trillion annually in each fiscal year since 2009, and deficits are projected to continue. Over time, unsustainable deficits can lead to reduced savings for investment, higher interest rates, and higher levels of inflation. Restoring fiscal balance would require spending reductions, revenue increases, or some combination of the two. Policymakers have considered a number of options for raising additional federal revenues, including a carbon tax. A carbon tax could apply directly to carbon dioxide (CO2) and other greenhouse gas (GHG) emissions, or to the inputs (e.g., fossil fuels) that lead to the emissions. Unlike a tax on the energy content of each fuel, a carbon tax would vary with a fuel's carbon content, as there is a direct correlation between a fuel's carbon content and its CO2 emissions. This book examines the carbon tax debate as a deficit reduction tool with a focus on policy design considerations.

Chapter 1 – The federal budget deficit has exceeded $1 trillion annually in each fiscal year since 2009, and deficits are projected to continue. Over time, unsustainable deficits can lead to reduced savings for investment, higher interest rates, and higher levels of inflation. Restoring fiscal balance would require spending reductions, revenue increases, or some combination of the two.

Policymakers have considered a number of options for raising additional federal revenues, including a carbon tax. A carbon tax could apply directly to carbon dioxide (CO_2) and other greenhouse gas (GHG) emissions, or to the inputs (e.g., fossil fuels) that lead to the emissions. Unlike a tax on the energy content of each fuel (e.g., Btu tax), a carbon tax would vary with a fuel's carbon content, as there is a direct correlation between a fuel's carbon content and its CO_2 emissions.

Carbon taxes have been proposed for many years by economists and some Members of Congress, including in the 112th Congress. If Congress were to establish a carbon tax, policymakers would face several implementation decisions, including the point and rate of taxation. Although the point of taxation does not necessarily reveal who bears the cost of the tax, this decision involves trade-offs, such as comprehensiveness versus administrative complexity.

Several economic approaches could inform the debate over the tax rate. Congress could set a tax rate designed to accrue a specific amount of revenues. Some would recommend setting the tax rate based on estimated benefits associated with avoiding climate change impacts. Alternatively, Congress could set a tax rate based on the carbon prices estimated to meet a specific GHG emissions target.

Carbon tax revenues would vary greatly depending on the design features of the tax, as well as market factors that are difficult to predict. One study estimated that a tax rate of $20 per metric ton of CO_2 would generate approximately $88 billion in 2012, rising to $144 billion by 2020. The impact such an amount would have on budget deficits depends on which budget deficit projection is used. For example, this estimated revenue source would reduce the 10-year budget deficit by 50%, using the 2012 baseline projection of the Congressional Budget Office (CBO). However, under CBO's alternative fiscal scenario, the same carbon tax would reduce the 10-year budget deficit by about 12%.

When deciding how to allocate revenues, policymakers would encounter key trade-offs: minimizing the costs of the carbon tax to "society" overall versus alleviating the costs borne by subgroups in the U.S. population or specific domestic industries. Economic studies indicate that using carbon tax revenues to offset reductions in existing taxes—labor, income, and investment— could yield the greatest benefit to the economy overall. However, the approaches that yield the largest overall benefit often impose disproportionate costs on lower-income households.

In addition, carbon-intensive, trade-exposed industries may face a disproportionate impact within a unilateral carbon tax system. Policymakers could alleviate this burden through carbon tax revenue distribution or through a border adjustment mechanism. Both approaches may entail trade concerns.

Chapter 2 – Market-based mechanisms that limit greenhouse gas (GHG) emissions can be divided into two types: quantity control (e.g., cap-and-trade) and price control (e.g., carbon tax or fee). To some extent, a carbon tax and a cap-and-trade program would produce similar effects: Both are estimated to increase the price of fossil fuels, which would ultimately be borne by consumers, particularly households. Although there are multiple tools available to policymakers that could control GHG emissions—including existing statutory authorities—this report focuses on a carbon tax approach and how it compares to its more frequently discussed counterpart: cap-and-trade.

If policymakers had perfect information regarding the market, either a price (carbon tax) or quantity control (cap-and-trade system) instrument could be designed to achieve the same outcome. Because this market ideal does not exist, preference for a carbon tax or a cap-and-trade program ultimately depends on which variable one wants to control—emissions or costs. Although there are several design mechanisms that could blur the distinction, the gap between price control and quantity control can never be completely overcome.

A carbon tax has several potential advantages. With a fixed price ceiling on emissions (or their inputs—e.g., fossil fuels), a tax approach would not cause additional volatility in energy prices. A set price would provide industry with better information to guide investment decisions: e.g., efficiency improvements, equipment upgrades. Economists often highlight a relative economic efficiency advantage of a carbon tax, but this potential advantage rests on assumptions—about the expected costs and benefits of climate change mitigation—that are uncertain and controversial. Some contend that a carbon tax may provide implementation advantages: greater transparency, reduced administrative burden, and relative ease of modification.

The primary disadvantage of a carbon tax is that it would yield uncertain emission control. Some argue that the potential for irreversible climate change impacts necessitates the emissions certainty that is only available with a quantity-based instrument (e.g., cap-and-trade). Although it may present implementation challenges, policymakers could devise a tax program that allows some short-term emission fluctuations, while progressing toward a long-term emission reduction objective. Proponents argue that short-term emission fluctuations would be preferable to the price volatility that might be expected with a cap-and-trade system.

Although a carbon tax could possibly face more political obstacles than a cap-and-trade program, some of these obstacles may be based on misunderstandings of the differences between the two approaches or on assumptions that the tax would be set too low to be effective. Carbon tax proponents could possibly address these issues to some degree, but there remains considerable political momentum for a cap-and-trade program.

Chapter 3 – State and local governments in the United States are evaluating a wide range of policies to reduce carbon emissions, including, in some instances, carbon taxes. Carbon taxes have existed internationally for nearly 20 years.

This report reviews existing carbon tax policies both internationally and in the United States. It also analyzes carbon policy design and effectiveness. Table ES-1 summarizes the carbon tax rates, annual revenue collected, and revenue distribution methods of existing and proposed carbon tax policies.

While each carbon tax is designed to meet the needs of the implementing jurisdiction, the implementation of all carbon taxes involves design considerations such as which sectors to tax, where to set the tax rate, how to use tax revenues, what the impact will be on consumers, and how to ensure emissions reduction goals are achieved.

Carbon taxes are most commonly placed on gasoline, coal, and natural gas. There are large differences among tax rates across various jurisdictions. The highest rates are paid in Sweden ($105 per metric ton CO_2), while the lowest rates are paid in the Bay Area Air Quality Management District (BAAQMD) in California ($0.045 per metric ton CO_2). How jurisdictions set tax rates depends on what outcomes they are trying to achieve; higher rates can induce behavior change, while lower rates may be more appropriate to generate funds for carbon mitigation programs. Some jurisdictions, such as Sweden, allow industries to pay a reduced tax rate.

In: Carbon Taxes
Editor: Neil Viveiros

ISBN: 978-1-62618-148-9

Chapter 1

CARBON TAX: DEFICIT REDUCTION AND OTHER CONSIDERATIONS[*]

Jonathan L. Ramseur, Jane A. Leggett and Molly F. Sherlock

SUMMARY

The federal budget deficit has exceeded $1 trillion annually in each fiscal year since 2009, and deficits are projected to continue. Over time, unsustainable deficits can lead to reduced savings for investment, higher interest rates, and higher levels of inflation. Restoring fiscal balance would require spending reductions, revenue increases, or some combination of the two.

Policymakers have considered a number of options for raising additional federal revenues, including a carbon tax. A carbon tax could apply directly to carbon dioxide (CO_2) and other greenhouse gas (GHG) emissions, or to the inputs (e.g., fossil fuels) that lead to the emissions. Unlike a tax on the energy content of each fuel (e.g., Btu tax), a carbon tax would vary with a fuel's carbon content, as there is a direct correlation between a fuel's carbon content and its CO_2 emissions.

Carbon taxes have been proposed for many years by economists and some Members of Congress, including in the 112th Congress. If Congress were to establish a carbon tax, policymakers would face several

[*] This is an edited, reformatted and augmented version of Congressional Research Service, Publication No. R42731, dated September 17, 2012.

implementation decisions, including the point and rate of taxation. Although the point of taxation does not necessarily reveal who bears the cost of the tax, this decision involves trade-offs, such as comprehensiveness versus administrative complexity.

Several economic approaches could inform the debate over the tax rate. Congress could set a tax rate designed to accrue a specific amount of revenues. Some would recommend setting the tax rate based on estimated benefits associated with avoiding climate change impacts. Alternatively, Congress could set a tax rate based on the carbon prices estimated to meet a specific GHG emissions target.

Carbon tax revenues would vary greatly depending on the design features of the tax, as well as market factors that are difficult to predict. One study estimated that a tax rate of \$20 per metric ton of CO_2 would generate approximately \$88 billion in 2012, rising to \$144 billion by 2020. The impact such an amount would have on budget deficits depends on which budget deficit projection is used. For example, this estimated revenue source would reduce the 10-year budget deficit by 50%, using the 2012 baseline projection of the Congressional Budget Office (CBO). However, under CBO's alternative fiscal scenario, the same carbon tax would reduce the 10-year budget deficit by about 12%.

When deciding how to allocate revenues, policymakers would encounter key trade-offs: minimizing the costs of the carbon tax to "society" overall versus alleviating the costs borne by subgroups in the U.S. population or specific domestic industries. Economic studies indicate that using carbon tax revenues to offset reductions in existing taxes—labor, income, and investment— could yield the greatest benefit to the economy overall. However, the approaches that yield the largest overall benefit often impose disproportionate costs on lower-income households.

In addition, carbon-intensive, trade-exposed industries may face a disproportionate impact within a unilateral carbon tax system. Policymakers could alleviate this burden through carbon tax revenue distribution or through a border adjustment mechanism. Both approaches may entail trade concerns.

INTRODUCTION

The federal budget deficit has exceeded \$1 trillion annually in each fiscal year since 2009. In August 2012, the Congressional Budget Office (CBO) estimated an FY2012 budget deficit of \$1.1 trillion, or 7.3% of gross domestic product (GDP).[1] Under CBO's alternative fiscal scenario, which assumes continuation of many current policies, the deficit as a percentage of GDP will be above 5% and rising from 2021 onward. Budget deficits are projected to continue, as the current mix of federal fiscal policies is widely viewed as being unsustainable in the long term.[2] Over time, unsustainable deficits can have negative macroeconomic consequences, including reduced savings for investment, higher interest rates, and higher levels of inflation.[3] Restoring fiscal balance will require spending reductions, revenue increases, or some combination of the two.

Policymakers have considered a number of options for raising additional federal revenues (see the text box below). One potential option is a carbon tax. Several economists and policy analysts from across the political spectrum have expressed interest in a carbon tax mechanism in recent years.[4] As of the date of this report, Members have introduced two carbon tax bills in the 112th Congress.[5] Several carbon price systems were proposed in the 111th Congress—as identified in the Appendix—most frequently as an efficient means to stimulate greenhouse gas (GHG) emission reductions. In addition, some countries have levied carbon taxes (or something similar) for over 20 years.[6]

Options for Raising Federal Revenues

Congress could utilize one or more policy options to raise federal revenue. One frequently discussed option is to "broaden the tax base" by reforming or eliminating certain tax preferences. Some proponents of tax-base broadening suggest that any revenues raised should be used to reduce marginal rates, such that tax reform is revenue neutral.[7] Accordingly, tax-base broadening may have limited potential to generate additional federal revenues.[8] A second option for raising additional revenues is to increase other existing taxes. For example, Social Security reform might include increased payroll taxes.[9] Another option would be an increase in the motor fuel excise tax (e.g., the gas tax).[10] In addition, a number of alternative revenue sources could be used for deficit reduction, many of which are not widely used at the federal level. As one example, the United State could consider a value-added tax (VAT).[11]

Other policy considerations, including environmental concerns, may also lead to consideration of a carbon tax. GHG in the atmosphere trap radiation as heat, warming the Earth's surface and oceans. The key human-related GHG is carbon dioxide (CO_2), primarily generated through the combustion of fossil fuels: coal, oil, and natural gas.[12] Although fossil fuels have facilitated economic growth in the United States and around the world, fossil fuel combustion has inadvertently raised the atmospheric concentration of CO_2 by about 40% over the past 150 years.[13] Almost all climate scientists agree[14] that these CO_2 increases have contributed to a warmer climate today, and that, if they continue, will contribute to future climate change. Many economists have argued that current fossil fuel prices reflect a *market failure,* because GHG emissions from fossil fuels contribute to current and future climate change, yet fossil fuel prices do not reflect climate change-related costs. Market failures distort economic efficiency by affecting consumer behavior. For example, energy consumers may make choices that are not in society's best interest, consuming more than the optimal amount of GHG emitting fuels, if prices do not reflect climate change-related costs. Carbon taxes, or GHG fees, have been proposed as one means to correct such a market failure. Another option is a cap-and-trade program, which would attach a price to GHG emissions by limiting their generation.[15] To some extent, a carbon tax and cap-and-trade program would produce similar effects: Both would place a price on carbon, and both are estimated to increase the price of fossil fuels. Preference between the two approaches ultimately depends on which variable one prefers to control—GHG emissions or costs.[16] A tax based on carbon content of fuels is different from a tax based on energy content, such as a Btu tax (see the text box below). Placing an emissions fee on CO_2 (and possibly other GHG emissions) could stimulate lower emissions and spur innovation in new lower-emitting technologies. A tax or fee based approach would allow markets to determine the level of investment in lower-emissions technologies. In addition, carbon tax revenues could be used to support multiple objectives, such as deficit reduction, or to replace existing taxes (e.g., payroll, income).

The 1993 Btu Tax

The deficit reduction package proposed by President Clinton in 1993 would have levied a tax based on energy content, measured in British thermal units (Btu). The 1993 Btu tax proposal called for a levy of 25.7 cents per million Btu, with a surcharge of 34 cents/million Btu on petroleum. The goals of the 1993 Btu tax proposal were to promote energy conservation and raise revenue. At the time, the proposed tax would have generated a new revenue stream of about $30 billion per year. The proposal was met with strong opposition and was not enacted; Congress ultimately enacted an (approximately 5-cent per gallon) increase in the motor fuels taxes.

The first section of this report examines carbon tax design and implementation issues, including the point of taxation, the rate of taxation, and the distribution of tax revenue. The second section discusses several carbon tax policy considerations: revenue potential, economic efficiency, equity, and operability. The final section highlights key issues related to the use of carbon tax revenues. Specifically, how might addressing the regressivity of a carbon tax diminish its revenue-raising potential, and what is the potential for a carbon tax to contribute to deficit reduction goals?

DESIGN AND IMPLEMENTATION

When establishing a carbon tax, there are several implementation decisions to be considered. Key considerations include (1) the point of taxation—where to impose the tax and what to tax; and (2) the rate of taxation. Information on how a carbon tax might affect fossil fuel prices is also provided.

Point of Taxation

The point of taxation would influence which entities would be required to (1) make tax payments based on emissions or emission inputs (e.g., fossil fuels), (2) monitor emissions or emission inputs, and (3) maintain records of relevant activities and transactions. The point of taxation does not necessarily reveal who bears the cost of the tax, as the cost may be passed on to intermediate producers or consumers. (This will be discussed in a later section.)

GHG emissions are generated throughout the economy by millions of discrete sources: smokestacks, vehicle exhaust pipes, households, commercial buildings, livestock, etc. Although CO_2 is the primary GHG at many sources, some sources predominantly emit non-CO_2 GHGs, such as methane.[17] When determining which sources and which GHG gases to control through a tax, policymakers would need to balance the benefits of comprehensiveness with administrative complexity and costs. Applying a carbon tax at the points of emissions from all GHG sources would present enormous logistical challenges.

CO_2 emissions are fairly easily to verify from large stationary sources, such as power plants. For almost 20 years, measurement devices have been installed in smokestacks of large facilities, reporting electronic information to the U.S. Environmental Protection Agency. For smaller sources, CO_2 emissions are a straightforward and accurate calculation based on the carbon content of fossil fuels consumed. Other GHG emissions, such as methane, nitrous oxides, sulfur hexafluoride, and others, could be more difficult or less reliable to verify. Thus, administrative costs and non-compliance risks would likely increase with a broader scope of an emissions tax. However, limiting the tax scope could result in perverse effects, with sources potentially shifting processes, facility size, or location to avoid taxes.

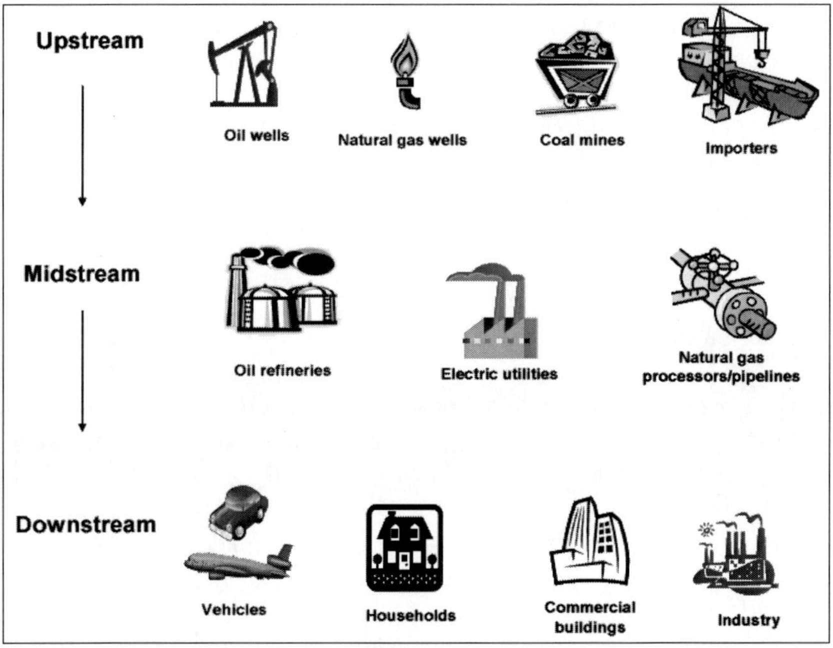

Source: Prepared by CRS.

Note: Electric utilities could be listed as either downstream entities—because they are direct sources of emissions—or midstream, because their emissions are tied to the electricity consumption of their customers, the further downstream consumers.

Figure 1. Illustration of Options for Points of Taxation within the Energy Production-to-Consumption Chain.

Policymakers may consider limiting the tax to sectors or sources that emit large percentages of the total U.S. GHG emissions or those that are relatively easy to measure. Or, the tax could be applied to reliable proxies for emissions in the production-to-emission chain. For example, the fossil fuel supply chain offers some options. As illustrated in Figure 1, potential emissions could be taxed at an "upstream" stage in that process, when the carbon-containing fossil fuel is first sold following production. Or, the point of taxation could, theoretically, be "downstream" where the pollution is released to the atmosphere.[18] Table 1 lists the top emission sources of six principal GHGs in the United States. These sources combine to account for approximately 95% of these gases (based on 2010 data).[19] Table 1 also provides potential points of taxation that would cover the emissions from these sources. For example, policymakers could address CO_2 emissions from fossil fuel combustion and non-energy uses by levying an "upstream"[20] carbon tax on fewer than 2,300 entities—in aggregate 80% of U.S. GHG emissions. Alternatively, policymakers could employ a "downstream" approach, applying a carbon tax at the point where the gas is released to the atmosphere. For example, the FY2008 Consolidated Appropriations Act (P.L. 110-161) requires all large sources to annually *report* their GHG emissions to the EPA's Greenhouse Gas Reporting Program (GHGRP).[21] The program covers between 85-90% of all U.S. GHG emissions from approximately 13,000 facilities.[22]

Table 1. Selected Sources of U.S. GHG Emissions and Potential Applications of a Carbon Tax

		Potential Carbon Tax Applications	
GHG Emission Source	Percentage of U.S. GHG Emissions (2010 data)	Entity	Number
CO_2 from fossil fuel combustion:	78.9	Coal mines[a]	1,257
- electricity generation		or	
- transportation		Coal-fired power plants[b]	641
- industrial		Power plantsc using imported coal[d]	26
		Petroleum refineries[e]	150
- commercial/residential		Petroleum importers[f]	220
		Natural gas processors[g]	530
		Natural gas importers[h]	45
N_2O from agricultural soils	3.0	Farms[i]	>2 million
CO_2 from non-energy use of fuels	1.8	Covered by the tax applied to fuels (above)[j]	
CH_4 from livestock (enteric fermentation)	2.1	Cattle operations[k]	967,440
CH_4 from landfills	1.6	Landfills[l]	1,800

Table 1. (Continued)

		Potential Carbon Tax Applications	
GHG Emission Source	Percentage of U.S. GHG Emissions (2010 data)	Entity	Number
HFCs from the substitution of ozone depleting substances	1.7	HFC manufacturers[m]	5
CH_4 from natural gas systems	3.1	Natural gas processors	530
CH_4 from coal mines	1.1	Coal minesn	1,257
CO_2 from iron/steel production	0.8	Raw steel production facilities;	116
		Integrated steel mills[o]	18
CO_2 from cement manufacturing	0.5	Cement plants[p]	118
CH_4 from manure management	0.8	Cattle operations;	967,440
		Swine operations[q]	65,640
Percentage of Total GHG Emissions	95.4		

Source: Prepared by CRS; GHG emission data from EPA, EPA, *Inventory of U.S. Greenhouse Gas Emissions and Sinks: 1990*-2010, April 2012; data for number of entities from multiple sources, cited in notes below.

a This figure accounts for mines in operation in 2010. EIA, *Coal Production and Number of Mines by State and Mine Type*, (2011).

b In 2006, there were 641 plants with at least one coal-fired generating unit (EIA, 860 Database).

c Number of plants comes from EIA database, Monthly Nonutility Fuel Receipts and Fuel Quality Data (Database 423). CRS was unable to determine the number of companies that act as coal importers, analogous to petroleum importers.

d In 2006, the United States imported approximately 36 million short tons of coal (EIA, *Quarterly Coal Report* (2008), table 4)—3.5% of the amount of coal consumed domestically in that year (EIA, *Annual Coal Report 2006* (2007), table 26). Coal imports have increased by more than 200% since 2002.

e This figure represents the number of "operable" refineries. EIA, *Refinery Capacity Report* (2008).

f EIA, Company Level Imports (as of November 2008). Note that some of these companies may import only crude oil, whose emissions would be covered by the tax at the domestic refineries. Thus, this figure represents an upper bound of petroleum product importers potentially subject to a carbon tax.

g EIA, Natural Gas Processing: The Crucial Link Between Natural Gas Production and Its Transportation to Market (2006).

h This includes pipelines and liquefied natural gas facilities. EIA, *About U.S. Natural Gas Pipelines* (as of September 2008).

i U.S. Department of Agriculture, *Farms, Land in Farms, and Livestock Operations: 2007 Summary* (2008). The resource defines a farm as "any place from which $1,000 or more of agricultural products were produced and sold, or normally would have been sold, during the year."

j Fossil fuels are used for a wide range of non-energy purposes. EPA estimates that of the total carbon consumed for non-energy purposes, approximately 62% is stored in products, and not released to the atmosphere (EPA, *Inventory of U.S. Greenhouse Gas Emissions and Sinks: 1990-2006* (April 2008), tables 3-14 and 3-15). The 2% value in Table 4 represents the emissions. In an upstream carbon tax system, fuels would be taxed before they are used. Congress could choose to consider providing tax credits for the amount of carbon stored in products.

k U.S. Department of Agriculture, Farms, Land in Farms, and Livestock Operations: 2007 Summary (2008).

l EPA, Inventory of U.S. Greenhouse Gas Emissions and Sinks: 1990-2006 (April 2008), citing BioCycle, 15t[h] Annual BioCycle Nationwide Survey: The State of Garbage in America (2006).

m Intergovernmental Panel on Climate Change, Safeguarding the Ozone Layer and the Global Climate System, Issues related to Hydrofluorocarbons and Perfluorocarbons (2005), Figure 11.1.

n Methane from underground mines, which accounts for about 61% of coal mine methane, is removed through ventilation systems for safety reasons. These emissions would be easier to monitor under a carbon tax than aboveground coal mine methane emissions.

o Data from U.S. Geological Survey, Mineral Commodity Summary, Iron and Steel Production (2008), at http://minerals.usgs.gov/minerals/pubs /commodity /iron_&_steel/.

p Cement manufacturing information from Portland Cement Association, at http://www.cement.org/basics/ cementindustry.asp.

q U.S. Department of Agriculture, *Farms, Land in Farms, and Livestock Operations: 2007 Summary* (2008). Other animals—chickens, horses, and sheep—contribute approximately 10% of the total emissions from manure (EPA, *Inventory of U.S. Greenhouse Gas Emissions and Sinks: 1990-2006* (April 2008), table 6-6).

Rate of Taxation

Although setting the carbon tax rate would likely involve political considerations, several economic approaches could be used to inform the debate. As a public finance instrument, the tax rate could be based on the estimated revenues needed to reduce or eliminate projected budget deficits.[23] This approach would be challenging, because estimates of future budget deficits (and thus the revenues needed to address them) are "inherently uncertain."[24] Moreover, a carbon tax would have some impact on the overall economy (and thus the overall tax base) that would make such a calculation more difficult.

Alternatively, the tax rate could be tied to climate change objectives. For example, Congress could set a tax rate based on the estimated benefits associated with avoiding climate change impacts. Some economists would say that the optimal level of a carbon tax would be at the "marginal cost of climate change," which is the incremental cost of damages of one more ton of emissions. This is sometimes called the Social Cost of Carbon (SCC).[25] The rate could include, as well, the cost of incremental damages of ocean acidification and other possible effects associated with carbon emissions (e.g., related air pollution).[26]

Estimates of the risks of GHG emissions are uncertain and cover a wide range. Analysts must place monetary values on goods and services that may be difficult or controversial to estimate, such as human health/life, water supplies, agricultural production, recreational activities. In addition, the element of time particularly complicates the valuation. The factor of time would demand a consideration of what global society should be willing to pay *now* to avoid *future* damages due to additional emissions generated *today*.[27] In short, basing the tax rate on a precise estimate of GHG emission-related risks would present extraordinary challenges.[28]

An alternative climate change approach would involve basing the tax rate on an estimate of the carbon price needed to meet a specific GHG emissions target. Such estimates, though relatively more certain than the SCC, are similarly based on multiple assumptions. Accordingly, estimates of carbon prices in hypothetical carbon reduction schemes have varied dramatically. For example, multiple parties prepared such estimates during the development of H.R. 2454 (in the 111th Congress), which intended to reduce GHG emissions to 17% below their 2005 levels by 2020.[29]

The emission allowance price estimates ranged from $16/metric tons of CO_2 equivalent (mtCO_2e) to $49/t$CO_2$e in 2015, with the estimated range increasing over time.[30] Moreover, it could be challenging to reach political agreement on the GHG emissions target sought as well as the tax rate that would achieve it.

Given the uncertainty in the above estimates and a desire to avoid "shocking" the economy with a sudden change in tax policy, some have proposed starting a carbon tax with a rate that is initially set at a low rate, with that rate rising annually as announced for a fixed period or indefinitely. This approach would have several potential advantages: it is simple to explain and understand; it provides predictability to investors and consumers; and it allows policymakers to hedge[31] against the risks of carbon emissions.

Carbon Tax Effects on Fossil Fuel Prices

Different fossil fuels generate different amounts of CO_2 emissions per unit of energy. Therefore, a carbon tax, which is based on CO_2 emissions, would levy a higher charge (per unit of energy) on some fuels than others. As indicated in Table 2, coal generates approximately 80% more CO_2 emissions per unit of energy than natural gas, and approximately 28% more emissions per energy than crude oil. These differences in emissions intensity would lead to different tax rates per unit of energy across different fuels in a carbon tax regime.

Table 2. CO_2 Emissions Per Unit of Energy for Fossil Fuels

Fossil Fuel	CO_2 Emissions Per Unit of Energy (million metric tons/quadrillion BTU)
Coal	96
Crude oil	75
Natural gas	53

Source: Prepared by CRS, based on Energy Information Administration (EIA), "Emissions Factors and Global Warming Potentials," updated January 2011, at http://www.eia.gov/oiaf/ 1605/emission_factors.html.

Notes: Coal emissions intensity values vary by type of coal, from 93-104 million metric tons/quadrillion BTU. The value above is for coal use in the electric power sector. The natural gas value above is the weighted national average of all uses, as prepared by EIA.

Table 3 includes estimates of the tax levied on fossil fuels and motor gasoline at different carbon tax rates. The change in price that consumers would see would likely not be the same as the carbon tax. Carbon taxes could affect fuel prices in complex ways. For example, energy consumers, to the extent possible, would likely shift their preferences to less expensive fuels, and the underlying prices of the fuels would change.

Table 3. Estimated Taxes Levied on Fossil Fuels and Motor Gasoline Based on Selected Carbon Tax Rates

Carbon Tax Ratea	Coal	Crude Oil	Natural Gas	Motor Gasoline
\$5/ $mtCO_2$	\$9.50/ short ton	\$2.15/ barrel	\$0.25/ mcfb	\$0.05/ gallon
\$15/ $mtCO_2$	\$28.50/ short ton	\$6.45/ barrel	\$0.75/ mcf	\$0.15/ gallon
\$25/ $mtCO_2$	\$47.50/ short ton	\$10.75/ barrel	\$1.25/ mcf	\$0.25/ gallon
\$50/ $mtCO_2$	\$95.00/ short ton	\$21.50/ barrel	\$2.50/ mcf	\$0.50/ gallon
	CO_2 Emissions Intensities Used in Above Comparison			
Metric tons of CO_2 ($mtCO_2$) per unit of fuel	1.9 $mtCO_2$/ short tonc	0.43 $mtCO_2$/ barreld	0.05 $mtCO_2$/ mcfe	0.01 $mtCO_2$/ gallonf
Recent Market Prices for Each Fuel Type: Average Price Between 2000-2010 (2005 dollars)				
	Coal Type (\$/short ton)[g]	Production Location(\$/barrel) Domestic first purchase price:h \$47	Economic Sector (\$/mcf) Residential: \$11	Unleaded Regular (\$/gallon): \$2.14
	Bituminous coal: \$37	Import landed costs:i \$48	Commercial: \$10	
	Sub-bituminous: \$9		Industrial: \$6	
	Lignite: \$14		Transportation: \$8	
	Anthracite: \$48		Electric power: \$6	

Source: Prepared by CRS. CRS calculated the estimated taxes for each fuel by multiplying a carbon tax rate by the CO_2 emissions intensities for each fuel. CRS generated these inputs (the last row of the table) from the CO_2 coefficients (i.e., CO_2 emissions per quadrillion BTU) and thermal conversion factors (i.e., BTU per fuel unit) for each fuel. CO_2 coefficients are from EIA, "Emissions Factors and Global Warming Potentials," updated January 2011, at http://www.eia.gov/oiaf/1605/emission_factors.html; thermal conversion factors from EIA, *Annual Energy Review 2010*, October 2011, Appendices A1 (motor gasoline), A2 (crude oil), A4 (natural gas), and A5 (coal).

Fuel prices: Coal prices from EIA, *Annual Energy Review 2010*, Table 7.9, October 2011. Crude oil prices from Table 5.18 (domestic) and Table 5.19 (imported). CRS converted the nominal dollar values in Table 5.19 to \$2005, using the GDP implicit price deflator provided in Appendix D of the *Annual Energy Review 2010*. Natural gas prices from Table 6.8. Gasoline prices from Table 5.24.

a The tax rates in the table were selected for comparison purposes. The initial carbon tax rates in recent legislative proposals fall within the range identified above. Most proposals include annual rate increases, some of which generating tax rates that would approach the upper end of the range after several years.

b MCF is thousand cubic feet.

c This value represents the CO_2 coefficient for coal (electric power sector) and the thermal conversion factor for coal consumption from the electric power sector (2010).

d This value represents the CO_2 coefficient for crude oil and the thermal conversion factor for "unfinished oil."

e This value represents CO_2 coefficient for natural gas ("weighted national average") and the thermal conversion factor for end-use sectors (2010).

f This value represents the CO_2 coefficient for "motor gasoline" and the thermal conversion factor for motor gasoline (conventional).

g Bituminous and Subbituminous coal, in aggregate, accounted for 93% of U.S. coal production in 2010 (EIA, *Annual Energy Review 2010*, Table 7.2, October 2011).

h EIA defines this price as "the price for domestic crude oil reported by the company that owns the crude oil the first time it is removed from the lease boundary." (*Annual Energy Review 2010*, Glossary).

i EIA defines this price as "Crude Oil Landed Cost: The price of crude oil at the port of discharge, including charges associated with purchasing, transporting, and insuring a cargo from the purchase point to the port of discharge. The cost does not include charges incurred at the discharge port (e.g., import tariffs or fees, wharfage charges, and demurrage)." (*Annual Energy Review 2010*, Glossary).

Actual price increases that result from the illustrative carbon taxes in Table 3 would depend on whether

- a carbon tax is applied at the beginning of the production process ("upstream") to fossil fuels (Figure 1);
- the price impacts are passed through to end-users and not absorbed by upstream energy producers or midstream entities (such as retailers); and
- consumers modify their behavior in the marketplace—energy conservation, fuel substitution, etc.

FRAMEWORK FOR EVALUATION

Policymakers may consider a carbon tax based on a number of policy motivations and potential benefits, including

- increased federal revenues,
- reduced federal budget deficits,
- reduced GHG emissions, climate change, and ocean acidification,[32]
- enhanced energy security,
- improved economic efficiency, and/or
- reduced tax rates from other revenue sources.

In recent years in the United States, most carbon tax proposals have aimed primarily to discourage GHG emissions.[33] In these proposals, a tax or fee would be levied on GHG emissions or the inputs (i.e., fossil fuels) that lead to emissions. The tax or fee raises the price of the emission-generating products, and therefore can, on the one hand, motivate suppliers to reduce the emissions involved in making the product and, on the other hand, encourage consumers to buy less of the product. This policy instrument sets the tax (cost) per unit of emissions and relies on private decision-makers to find the most efficient means to reduce the emissions. A carbon tax is one of several "market mechanisms" that relies on the efficiencies of markets to maximize cost-effectiveness.

Though this policy instrument is commonly called a "carbon tax," its application could be broader than this term suggests. First, the policy may apply not just to CO_2 emissions, but also to multiple GHGs (e.g., methane or sulfur hexafluoride), including some that may have no molecular carbon.[34]

Second, if the levy's primary purpose were to charge those who use the atmosphere to absorb the full cost of their GHG emissions, the levy might instead be considered a "user fee."[35] Regardless, this report uses the term "carbon tax," because this report focuses on the policy instrument's potential to raise revenues and reduce the federal budget deficit. There are a number of issues to be considered when evaluating tax policy proposals.[36] The following sections analyze a generic carbon tax option using the criteria listed below:

- *adequacy*—the ability to generate a desired amount of revenues;
- *economic efficiency*—the potential to enhance or diminish the productivity of the U.S. economy;
- *equity*—the subjective determination of a proposal's fairness;
- *operability*—the combination of multiple factors, including administrative ease, transparency, avoidance of perverse outcomes, and consistency with federal and international norms and standards;[37] and
- *political feasibility*—the likelihood of enactment given a tax's visibility to the public and public opinion, differential regional implications, contribution to deficit reduction or other objectives, pledges made by some lawmakers not to raise taxes, etc.[38]

Adequacy—The Potential to Generate Revenues

The revenues that would be generated under a carbon tax vary greatly depending on the design features of the tax, namely the tax scope (i.e., base) and rate, as well as such independent factors as prices in global energy markets. Several recent proposals to price or tax carbon, where revenue estimates are available, are presented below. In 2011, the Congressional Budget Office (CBO) evaluated a hypothetical cap-and-trade program in which CO_2 emission allowances (i.e., permits to emit one metric ton of CO_2 emissions) would be sold at auction and traded in a carbon market. The auction revenues generated in this program would be analogous to tax revenues under a carbon tax system. CBO estimated the allowance price, which, under an actual program, would be determined by market forces, would begin at $20 per metric ton of CO_2 ($mtCO_2$) in 2012 and increase 5.6% annually. This allowance price estimate, and its projected annual increase, is akin to a prescribed carbon tax rate. CBO estimated that such a regime would raise $1.2 trillion over the 2012 to 2021 budget window (Table 4).

Table 4. Estimated Revenues from a CBO CO_2 Emissions Pricing Model $20/mtCO₂ in 2012, increasing 5.6% annually (billions of dollars)

	2012	2013	2014	2015	2016	2017	2018	2019	2020	2021	2012-2016	2012-2021
Revenues	88	93	99	105	112	119	127	136	144	154	499	1,17

Source: Congressional Budget Office, *Reducing the Deficit: Spending and Revenue Options*, Washington, DC, March 2011, pp. 205-206.

Notes: The CBO assessment does not precisely describe the scope of its carbon tax scenario (i.e., which industries would be covered). Also, estimated revenues reflect reductions in income and payroll tax revenues that would result from the effects of a charge for emitting carbon.

As a point of reference for the above carbon tax revenue estimates, **Figure 2** identifies the major sources of federal receipts in FY2011. As the figure illustrates, the range of potential carbon tax revenues is comparable (at least in the early years of the carbon tax system) to the revenues collected from current federal excise taxes. Nearly half of all federal excise tax receipts come from taxes on gasoline and diesel fuels.[39]

Legislation introduced in the 112th Congress, the Save Our Climate Act of 2011 (H.R. 3242), would establish a carbon tax on domestic and imported fossil fuels, as well as the carbon content of biomass, municipal solid waste, and any organic material used as fuel. The tax rate would begin at $10 per short ton[40] of CO_2 emissions (tCO_2), increasing by $10 annually until total U.S. CO_2 emissions are 20% or less of CO_2 emissions in 1990. The bill states that it would reduce the federal budget deficit by a $480 billion over ten years.

Other revenue estimates have been made in recent years of carbon taxes considered or included in a number of deficit and debt reduction proposals.[41] As one example, in 2010, the bipartisan Debt Reduction Task Force (Domenici-Rivlin) considered, but did not ultimately recommend, a tax on CO_2 emissions. The report estimated that a tax of $23 per ton of CO_2 emissions starting in 2018, increasing 5.8% annually, would raise approximately $1.1 trillion in cumulative revenues through 2025.[42]

In addition to a carbon tax design, a 2012 Resources for the Future (RFF) study[43] identified several factors—electricity demand and natural gas prices—that could influence carbon tax revenue. The study found that the magnitude of these factors' influence increases as the tax rate increases. Figure 3 illustrates the RFF study findings.

In Figure 3, RFF compares tax revenue estimates using assumptions from EIA's 2009 and 2011 *Annual Energy Outlook* (AEO) publications. According to the study, the AEO assumptions regarding future electricity demand and fuel prices, namely natural gas, varied dramatically between the 2009 and 2011 versions. These variances lead to different estimates of potential carbon tax revenue. For example, as illustrated in Figure 3, estimated revenues in 2020 from a tax rate of \$25/mtCO_2 would vary across scenarios by about \$10 billion. A \$40/mtCO_2 tax rate would generate an estimated revenue range across scenarios of almost \$20 billion in 2020.

Further, Figure 3 highlights a key result of a carbon tax that may affect its ability to provide a reliable source of revenue. Note that the tax revenues *in the electricity sector* begin to decline around 2030 with the \$40/mtCO_2 tax rate. The study concluded this decline is due to a diminishing reliance on fossil fuels and relatively greater reliance on low-carbon energy sources, such as renewables and nuclear.

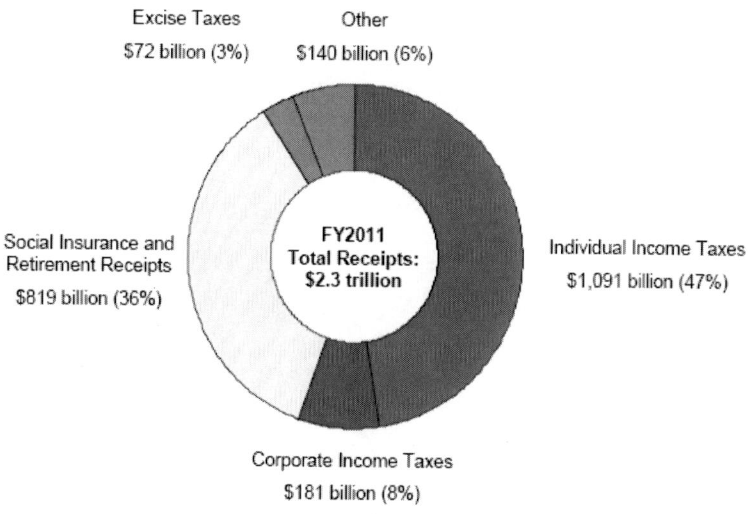

Source: Prepared by CRS with data from Office of Management and Budget, Historical Tables 2.1 and 2.2, at http://www.whitehouse.gov/omb/budget/historicals.

Figure 2. FY2011 Federal Receipts by Source.

The projected scenario in Figure 3 highlights a fundamental revenue adequacy concern associated with a carbon tax approach: if a primary goal of a carbon tax is to reduce its own tax base (i.e., carbon emissions), would this tax provide a reliable stream of revenues over time? With revenue reliability as a primary goal, governments impose taxes on activities to which producers and consumers are not very price-sensitive. In other words, the imposed taxes lead to minimal changes in market behavior. Researchers debate how much producers and consumers would respond to a tax on GHG emissions by reducing emissions or switching to goods and services that generate fewer emissions (because they embody lower taxes). With a higher degree of responsiveness, either revenues would decline or the tax rate per ton of emissions would need to increase over time to maintain the revenue stream (assuming that maintaining certain revenue stream is determined to be a policy objective).

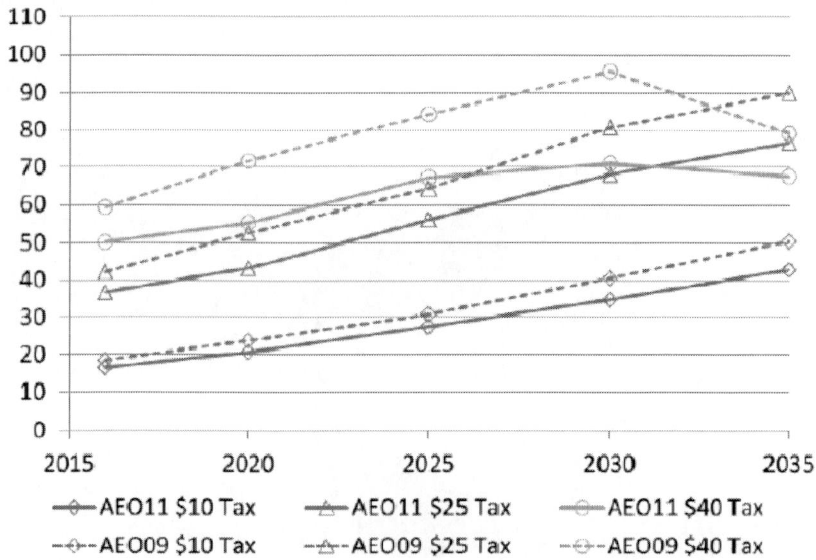

Source: Karen Palmer et al, *The Variability of Potential Revenue from a Tax on Carbon*, Resources for the Future, May 2012.

Notes: AEO09 and AEO11 refer to EIA's 2009 and 2011 *Annual Energy Outlook*. The carbon taxes are levied on metric tons of CO_2 emissions ($mtCO_2$), with rates starting in 2016 and growing 5% annually.

Figure 3. Annual Carbon Tax Revenues *in the Electricity Sector;* Billions of 2009$.

Economic Efficiency

Economic theory suggests that a carbon tax could improve the efficiency of the economy by at least three means:[44]

- First, markets could produce a more optimal mix of goods and services if the costs of emitting GHG while manufacturing products, providing services, or using goods were "internalized" into market prices; producers and consumers would more fully respond to the full costs of their decisions, resulting in a more economically efficient outcome;
- Second, a tax on an activity that yields pollutants, such as GHG emissions,[45] would discourage the polluting activity and therefore could be more efficient than an alternative tax that yields the same revenues but discourages a beneficial activity (e.g., investment); and
- Third, adding a smaller tax on a new activity, such as potential carbon emissions, could be less distortionary to production and consumption than increasing the tax rate on currently taxed activities.[46]

The difficulty associated with setting an optimal carbon tax rate makes achieving potential efficiency gains challenging. Further, a stand-alone carbon tax (e.g., one where revenues were not used to offset other taxes) could have efficiency costs, as a carbon tax would increase existing distortions on inputs in the production process. The difficulty of implementing an efficiency-enhancing carbon tax is compounded in a tax system that already contains a number of distortionary taxes.[47]

Many Taxes Have Distortionary Effects[48]

Many people are concerned about several potentially "distortionary" effects of taxes on society. Many economists consider that taxes take money away from people and change the relative prices of goods and services compared to what they would be without taxes; thus, taxes are considered "distortionary" by altering free market decisions. To minimize distortions (and generate a reliable stream of revenues), governments often prefer to tax products or activities for which consumer demands are relatively insensitive to price increases.[49] Some of these taxed items are recognized as essential or socially desirable, such as income, employment, and investment. Taxes may also be "distortionary" because they discourage a taxed but beneficial activity. For example, many economists argue that payroll and income taxes are distortionary because they discourage employment and investment.[50] If these taxes were reduced, the incentives to hire/work and invest, they argue, would be greater.

Also, as tax rates (i.e., the tax per unit of a product or service) increase, they become *more* distortionary. That is, higher tax rates have greater potential to shift choices away from optimal "free market" outcomes. As an example, consumers may respond proportionately more strongly to a $1.00 tax per gallon on gasoline than a $0.20 tax per gallon (in other words, the efficiency losses associated with a tax increase exponentially with the tax rate). Thus, applying lower tax rates to broader tax bases can help minimize inefficiencies.[51]

Taxes May Correct Market Failures

Not all results from economic activity are considered desirable—pollution being one example. When producers or consumers discharge pollution—including GHG emissions—to another person's private property or a publicly shared resource—such as the atmosphere—without paying to do so, they are not paying for the full cost of a product or activity. Economists would describe this outcome as a "market failure," because the costs associated with GHG emissions are not captured in the economic decision process. Economists contend that levying a charge on GHG emission would be an economically efficient way to correct the failure.[52] For example, in terms of environmental policy, fossil fuel prices do not reflect the costs—related to climate change and ocean acidification damages—associated with the GHG emissions. A pollution discharge fee could internalize these external costs into market prices.

A primary argument in favor of a carbon tax is that it would, in theory, increase the efficiency of markets by discouraging "bad" activities. A carbon tax would discourage pollution that imposes costs on others who do not necessarily benefit from the polluting activity. These may include future generations that bear the dislocations of climate change, or fishery sectors in developing countries that experience lower yields in acidified oceans.

A carbon tax would encourage energy consumers—for example, power plants, industry, households, etc.—to (1) switch to less carbon-intensive fuels; (2) use less energy or use energy more efficiently; and (3) prefer products or services that are lower-priced by virtue of incorporating less emission tax. Each of these activities would reduce GHG emissions compared to a business-as-usual track and could improve economic efficiency.

An Economically Efficient Carbon Tax Rate[53]

To improve efficiency, a carbon tax would need to be set at the "right" level. According to basic economic theory, the "right" level would be one that equilibrates the tax rate to the incremental harm, now and in the future, that a ton of GHG emissions imposes. To find this tax rate precisely, one would need to know the marginal costs of climate change and ocean acidification.

However, estimating the effects and placing a monetary value on them is both controversial and reliable only over a wide range. Hence, no consensus is likely to emerge regarding a precisely "right" carbon tax rate.[54] An analytical or political estimate is feasible but may not be the most economically efficient outcome. Further, setting carbon tax rates based on revenue needs for deficit reduction, or some other purpose, would not necessarily result in setting a rate that is proportional to the incremental harm of GHG emissions.

Equity

The "equity" or "fairness" of taxes can be evaluated by looking at how different parties are affected. How the burden of a tax is ultimately divided between different parties is described as the economic incidence of the tax.[55] To evaluate the subjective concepts of "equity" and "fairness" in tax policy, economists often examine two different types of equity: vertical and horizontal. In addition, some economists consider individual and generational equity. A complete policy analysis might consider the potential trade-off between economic efficiency and equity.[56] These four elements of tax equity are discussed below.

Vertical Equity

The notion of vertical equity suggests that those with a greater ability to pay should contribute more. Without some of tax revenue redistribution, carbon taxes are generally considered to be "regressive," because lower-income households generally spend a higher percentage of their income on energy-related goods and services than do higher-income households.[57] The actual or perceived regressivity of carbon or fuel taxes can have a strong influence on the political feasibility of the instrument.[58] For example, the coalition of advocates for low-income people and opponents of energy taxes arguably contributed to the failure of President Clinton's 1993 "Btu Tax" proposal (see the text box above), which was introduced as part of a deficit reduction package.

Particular carbon tax design elements could reduce its regressivity. For example, carbon tax revenues could be used to reduce other taxes (e.g., income) or be redistributed to lower-income households through a variety of funding mechanisms (see the section "Alternative Uses for Carbon Tax Revenues" below).

Horizontal Equity

Horizontal equity examines whether potential tax-payers with similar characteristics would receive equivalent tax treatment. Questions of horizontal equity may arise if particular industries or economic sectors that predominately emit non-CO_2 GHG emissions (e.g., methane) were exempted from the carbon tax regime, while industries or sectors of comparable size were included based on their CO_2 emissions.[59]

Individual Equity

Some people consider taxes unfair when they involve government interference in private transactions, including freedom to use one's resources as one chooses. This is sometimes considered a violation of "individual equity." [60] To minimize infringement of individual equity, some would argue that taxes should be levied on, and commensurate with, the benefit an individual receives from the taxed activity, or "benefit taxation."[61] (Proponents contend that not to violate individual equity would require that the tax be voluntary.) Some might counter that a carbon tax is a kind of "benefit taxation," levied on the benefit the carbon source receives by being allowed to discharge the pollution to the atmosphere.

Generational Equity

Are the burdens of taxation and benefits from governmental spending fairly distributed across generations? This is a particular concern in the context of the federal deficit and GHG-induced climate change, both of which would likely shift costs from the current generation to subsequent generations. At first glance, a carbon tax could potentially support generational equity by helping to slow or reduce GHG-induced climate change and by potentially reducing the deficit (depending on the fate of tax revenues).

However, an assessment of generational equity is complex. Some may argue that a carbon pricing mechanism would reduce the wealth of the current generation, consequently reducing the productive capacities of future generations. Moreover, some may argue that, instead of a carbon price approach, increased investment in technology would yield greater benefits for future generations. Others may counter that increased investments today may result in increased consumption, providing minimal benefits for future generations and foregoing the opportunity to address climate change impacts. Such an outcome would disproportionately burden future generations.

Operability

Some tax options may be elegant in terms of economic theory but, in practice, present considerable logistical challenges. This section identifies some of the key administrative issues that might arise when implementing a carbon tax.

Administrative Ease

In previous debates over carbon taxes, which largely focused on discouraging GHG emissions, many proponents asserted that carbon taxes would be administratively simple compared to the principal emissions control alternatives—notably, a cap-and-trade program[62] or emissions performance standards. As with other comparisons, the relative advantage of a carbon tax would depend on the designs of the instrument alternatives under scrutiny.

Any new kind of tax would require new administrative systems, but a carbon tax system could potentially take advantage of existing frameworks. A well-developed administrative structure for collecting taxes already exists in the United States. In addition, approximately 13,000 facilities report annual GHG emissions to EPA.[63] Therefore, EPA already collects carbon tax emissions data covering about 90% of all U.S. GHG emissions. However, a tax program would presumably be administered by the Department of the Treasury. Transferring data should not be technically difficult, but broader cooperation across agencies would likely be necessary to share respective expertise and to promote compliance.

Consistency with Federal and International Norms and Standards

Policymakers might consider the degree to which a carbon tax is consistent with other policies that support other objectives: pollution control, energy affordability, national security. Superficially, carbon taxes would seem to be consistent with other GHG emission control standards, but would conflict with incentives to make fossil fuels more affordable.

In terms of linking a U.S. GHG emission reduction scheme with international efforts, a carbon tax may be at a disadvantage, because the most prominent international activity, namely the European Union's Emission Trading Scheme,[64] currently involves a cap-and-trade approach.

Potential Perverse Effects

Some carbon pricing policies may yield unexpected and unwelcome results. For carbon taxes, one such "perverse effect" would be the potential to increase the cost of business activity in the United States. Over time, this could result in reduced revenue from both the carbon tax and other taxes.[65] This consideration has led some other countries to exempt energy-intensive manufacturers in specific or strategic sectors from energy or carbon taxes.[66]

In addition, a carbon tax could lead to emissions "leakage," if businesses moved operations overseas to avoid the tax. This outcome would depend on a number of factors, however, including relative fuel mixes, efficiencies of power production and transportation, border taxes, etc.

Transparency

Tax policies that are transparent may be more likely viewed as fair, as taxes being paid are understood by those responsible for paying the tax. In recent years, many have grown more skeptical of complex financial structures.[67] Potential complexity and opaqueness of a cap-andtrade program was one factor that led some advocates of GHG control policies to prefer carbon taxes as a policy instrument.

While a carbon tax could be designed to be simple and transparent, Congress could also choose to establish a carbon tax framework that rivals the complexity of a cap-and-trade program. For instance, policymakers could provide subsidies or exemptions to the fossil fuel industry or certain consumers (e.g., agricultural users) or enact incentives for producing, processing, and exporting fossil fuels. In addition, policymakers could allow for tax credits for carbon sequestration projects, similar to carbon offsets in a cap-and-trade regime.[68] As with carbon offsets in a capand-trade program, this would require a further level of administrative responsibilities, and potentially weaken the program.[69]

Ironically, transparency, particularly in regards to costs, could be a political liability for a carbon tax. Although both a carbon tax and a cap-and-trade program would impose higher energy costs, the costs from a cap-and-trade program would be more difficult to estimate, because the market would determine the price of emission allowances (and thus the overall costs of the program).

Political Feasibility

Certain stakeholders are likely to exercise strong opposition to a carbon tax. These include energy-intensive manufacturers, farmers, and regional energy interests—especially those whose asset values may fall with expected impacts on profitability of owned or leased coal and oil resources. One may look to numerous reviews of the histories of both the Clinton Btu tax proposal and more recent cap-and-trade bills for lengthier views regarding political feasibility.[70]

CONTRIBUTION TO DEFICIT REDUCTION

The possible contribution of a carbon tax to deficit reduction would depend on the magnitude and scope of the carbon tax, various market factors (discussed above), and assumptions about the size of the deficit. In August 2012, CBO released updated budget projections for fiscal years 2012 to 2022. Under current law, CBO estimated the 10-year budget deficit at $2.3 trillion, or 1.1% of GDP.[71] However, using an alternative fiscal scenario,[72] CBO's projected a larger deficit—$10.0 trillion, or 4.9% of GDP.

Enacting the carbon tax options discussed in the previous section could reduce future budget deficits. As illustrated in Figure 4, a $20/mtCO$_2$ price on carbon (increasing by 5.6% annually) would have a considerable impact on budget deficits using CBO's August 2012 baseline projection.

- The 10-year budget deficit could be reduced from $2.3 trillion to $1.1 trillion, or from 1.1% to 0.5% of GDP.
- Overall, a $20/mtCO$_2$ price on carbon would reduce the 10-year budget deficit by more than 50%.

Under CBO's alternative fiscal scenario, the same carbon tax would have a smaller impact on budget deficits.

- The deficit would be reduced from $10.0 trillion to $8.8 trillion, or from 4.9% to 4.4% of GDP.
- Overall, a $20/mtCO$_2$ price on carbon would reduce the 10-year budget deficit by about 12%.

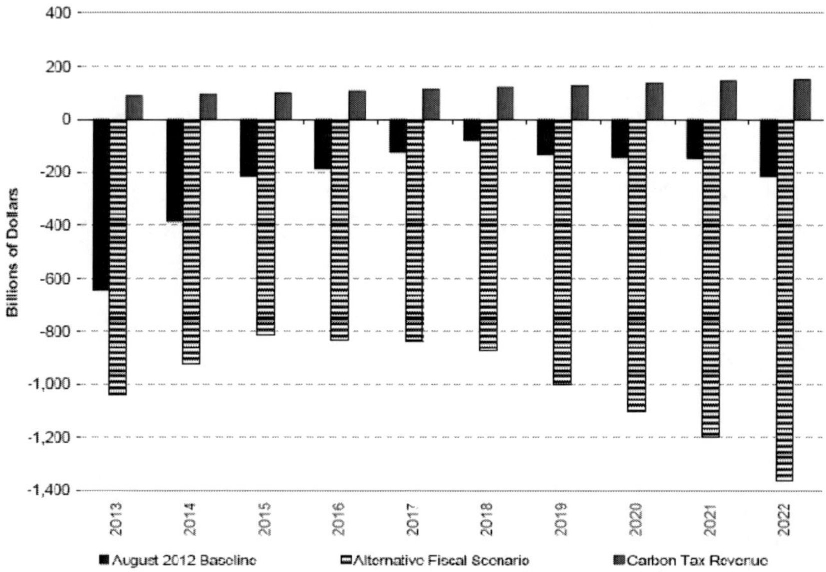

Source: Prepared by CRS. Carbon tax revenue estimates from CBO, *Reducing the Deficit: Spending and Revenue Options*, Washington, DC, March 2011, pp. 205-206; Budget deficit estimates from CBO, *An Update to the Budget and Economic Outlook: Fiscal Years 2012 to 2022*, August, 2012.

Notes: The 2012 baseline estimates are based on current law. The alternative scenario assumes that (1) most expiring tax provisions are extended and the Alternative Minimum Tax (AMT) is adjusted for inflation, (2) Medicare's payment rates for physicians' are held constant at current levels, and (3) that the automatic spending reductions required by the Budget Control Act (BCA) do not occur. Revenues from the carbon tax are assumed to start in FY2013.

Figure 4. CBO Estimated Revenues from a \$20/mtCO_2 Carbon Tax Compared to Two CBO Budget Deficit Projections; FY2013-FY2022.

Carbon tax proposals that raise less revenue would contribute less to deficit reduction. The Save Our Climate Act of 2011 (H.R. 3242) states that this legislation would reduce the deficit by \$480 billion over the 10-year budget window. Relative to the CBO's alternative fiscal scenario, the \$480 billion would reduce the budget deficit by 4.8%, from an estimated \$10.0 trillion to roughly \$9.5 trillion, or 4.9% to 5.7% of GDP.

The analysis above assumes that 100% of carbon tax revenue would be applied toward deficit reduction. The following section explores possible alternative uses for carbon tax revenues.

ALTERNATIVE USES FOR CARBON TAX REVENUES

Carbon tax revenues may be used to achieve a variety of policy goals. However, one revenue use necessarily forgoes the opportunity to apply that level of revenue to support other objectives, like deficit reduction. Therefore, in deciding how to allocate carbon tax revenues, policymakers would encounter trade-offs among objectives. Such trade-offs include

1. minimizing economy-wide costs resulting from a carbon tax;[73]
2. alleviating the costs borne by subgroups in the U.S. population, regions, economic sectors, and generations; and
3. supporting specific policy objectives, such as deficit reduction, climate change mitigation, energy efficiency, technological advances, domestic employment, or energy diversity.

A comprehensive discussion of alternative uses of carbon tax revenues and the involved tradeoffs is beyond the scope of this report.[74] Three possible options, which have received some attention in recent years, are discussed below.

Distribute Carbon Tax Revenues to Households

If Congress were to consider a carbon tax system, a key debate would likely involve the degree to which carbon tax revenues would be returned to households to alleviate the expected financial burden imposed by the carbon tax.[75] A 2007 study estimated that households and businesses that are end users would experience the vast majority (89%) of the private costs under a carbon pricing regime (the remaining 13% was attributed to coal, oil, and gas producers and fossil electricity generators).[76] Moreover, businesses—to the extent they were able—would likely pass through to household consumers some of their increased energy/electricity costs in the form of higher prices for their goods and services. Costs that are not passed through to consumers in the form of higher prices are borne by either labor, through reduced wages, or owners of capital, through reduced returns on investment.

Depending on how and to whom carbon tax revenues are distributed, lower-income households could face a disproportionate increase in tax burden. As illustrated in Table 5, a carbon tax "in isolation" (i.e., without revenue redistribution of some kind) could have a disproportionately greater impact on lower-income households than higher-income households—a regressive outcome.[77] As Table 5 illustrates, a carbon tax in isolation would reduce after-tax income for taxpayers in the lowest income deciles by 3.4%, while taxpayers in the highest income deciles would see their income fall by 0.8%.[78]

The remaining rows in Table 5 show how the household impacts would change if carbon tax revenues were rebated to households using different rebate mechanisms.[79]

- An equal lump-sum rebate of carbon tax revenues would be progressive, increasing incomes for those in the lowest income deciles relative to higher-income brackets.[80]
- A payroll tax rebate for workers, in this case, would reduce but not eliminate the regressivity of the carbon tax. Lower-income households without individuals in the workplace would not receive a rebate under this approach.[81]
- Adding a rebate for Social Security recipients to the payroll tax rebate would address the lower-income households with individuals not in the workplace, and further enhance the progressivity of this policy option.[82]

In this scenario, both a lump-sum carbon tax rebate and a carbon tax rebate that includes Social Security recipients increase the after-tax income of lower-income households while decreasing the after-tax income of higher-income households. Using carbon tax revenues to reduce payroll taxes would increase the after-tax income of middle-income households, while lower-income households would see their after-tax income fall.

Amongst the policies surveyed in Table 5, using carbon tax revenues to finance lump-sum rebates to households provides the most benefit to the lowest-income. While enhancing the progressivity of the overall tax system may be attractive from an equity perspective, lump-sum rebates to households offer limited potential for gains in overall economic efficiency. If carbon tax revenues are instead used to offset other distortionary taxes in the economy (i.e., payroll, income), the costs associated with a carbon tax may be reduced or eliminated.

Studies that examine the distribution of a carbon tax rely on a number of modeling assumptions. Metcalf (2007), for example, assumes that the costs of the carbon tax are related to energy expenditure patterns across income groups. An alternative approach is to look at the distribution of the carbon tax, assuming that the carbon tax reduces returns to factors of production: labor, capital, and fossil fuel resources.[83]

Under this approach, the regressivity of a carbon tax is reduced. If the burden of the carbon tax falls more heavily on owners of capital, as opposed to labor, the carbon tax affects higher income households that tend to derive more of their income from capital. Further, lower income households that receive a larger share of their income from transfer payments (e.g., social security), would be less affected by a carbon tax. While the "baseline" distribution of a carbon tax (i.e., the distribution of a carbon tax absent revenue recycling) differs under various modeling assumptions, the general conclusions regarding different policy choices for uses of carbon tax revenues often remain the same. Lump-sum redistribution of carbon tax revenues tends to be more progressive, while policies that reduce payroll taxes tend to be less regressive than those that reduce taxes on income or capital.[84]

Table 5. Distributional Effects of Carbon Tax with Different Applications of Carbon Tax Revenues

	Percentage Change in After-Tax Household Income by Income Bracket (1= Lowest, 10=Highest)									
Distributional Scenarios	1	2	3	4	5	6	7	8	9	10
Carbon Tax in Isolation (before revenue redistribution)	-3.4	-3.1	-2.4	-2.0	-1.8	-1.5	-1.4	-1.2	-1.1	-0.8
Carbon Tax with a Lump-Sum Distribution to Households	2.1	1.0	0.6	0.4	0.3	0.1	-0.1	-0.1	-0.2	-0.2
Carbon Tax with Payroll Tax Rebate for Workers	-0.7	-1.0	-0.2	0.1	0.1	0.3	0.2	0.2	0.0	0.0
Carbon Tax with Payroll Tax Rebate for Workers and Equivalent Rebate for Social Security Recipients	1.4	1.0	0.6	0.3	0.1	0.1	0.1	-0.1	-0.1	-0.2

Source: Prepared by CRS with data from the following: Gilbert Metcalf, *A Proposal for a U.S. Carbon Tax Swap: An Equitable Tax Reform to Address Global Climate Change* (2007), The Hamilton Project, Brookings Institution.

Note: These results do not account for consumers' behavioral responses to the carbon tax. Metcalf points out that the results provide a "reasonable first approximation" of the different impacts to household incomes (Metcalf (2008)).

Address Economy-Wide Costs

Some are concerned about potential economy-wide costs that a carbon tax would impose. Economy-wide costs, or macroeconomic costs, are often measured in terms of changes in projected gross domestic product (GDP) or another societal-scale metric, such as efficiency cost or welfare changes.

A tax on carbon would likely lead to increased energy costs, which could reduce GDP growth.[85] However, most measures of economic costs imposed by a carbon tax do not consider the climate change benefits or ancillary benefits[86] that a carbon price would provide. The ultimate economic effects would depend on a number of factors, including the magnitude, design, and use of revenues of the carbon tax.

Economic studies indicate that using carbon tax revenues to offset reductions in distortionary taxes—labor, income, and investment—would be the most economically efficient use of the revenues and yield the greatest benefit to the economy overall. Studies also conclude that using tax revenues to lower the federal deficit would yield an economy-wide benefit, because of the reduced need to impose distortionary taxes in the future.[87] But this benefit would be delayed and its realization assumes policymakers would, sometime in the future, address the deficit by raising taxes.

Using carbon tax revenues to offset other distortionary taxes is sometimes described as revenue recycling, a fiscal strategy that may yield a "double-dividend:"[88]

1. reduced GHG emissions, achieved through the new carbon price, and
2. reduced market distortions, achieved through lower taxes on desirable behavior.

According to a 2011 study,[89] economic literature[90] generally finds that revenue recycling may reduce the economy-wide costs imposed by a carbon tax intended to reduce GHG emissions, but may not eliminate them entirely.[91] However, the same 2011 study and a 2012 study both provide scenarios in which a carbon tax and revenue recycling would produce a net increase in Gross Domestic Product.[92] In other words, these studies found that, in certain situations, that the economic improvements gained by reducing existing distortionary taxes would outweigh the costs imposed by the new carbon tax.[93] In both studies, the potential benefits of reduced GHG emissions were not part of the calculation, largely because these estimates carry considerable uncertainty.

For example, the 2012 study[94] projected a net reduction in GDP and employment—a measure of economy-wide costs—when carbon tax revenues where used to (1) reduce the deficit, (2) provide lump-sum transfers to households, or (3) reduce payroll taxes. However, when carbon tax revenues were used to reduce marginal tax rates on capital income (e.g., the corporate tax), employment and GDP increased relative to the baseline. While using carbon tax revenues to reduce tax rates on capital may enhance economic efficiency, such policies would not offset the increased burden imposed by a carbon tax on low-income households.[95]

Assist Carbon-Intensive, Trade-Exposed Industries

Carbon-intensive, trade-exposed industries would likely face disproportionate impacts within a U.S. carbon tax system. This issue received considerable attention during the debate over H.R. 2454 in 2009 (the 111th Congress) and would likely receive similar attention if Congress were to consider establishing a carbon tax system. An industry's carbon intensity is a function of both direct CO_2 emissions from its manufacturing process (e.g., CO_2 from cement or steel production) and indirect CO_2 emissions from the inputs to the manufacturing process (e.g., electricity, natural gas). In general, trade-exposed industries are those that face considerable international competition. A carbon tax would present a particular challenge for these industries. Compared to other domestic industries, they would be less able to pass along the tax in the form of higher prices, because they may lose global market share (and jobs) to competitors in countries lacking comparable carbon policies.[96] To address these impacts, policymakers could distribute a portion of the carbon tax revenues to these industries. As one point of reference, under H.R. 2454 (111th Congress) such industries would have received approximately 15% of the emission allowance value—analogous to 15% of carbon tax revenue—through 2025, steadily decreasing to zero thereafter. At the time, some questioned whether this allotment was sufficient.[97] Regardless, the data and administrative resources necessary to implement such a program would be substantial. Alternatively, policymakers could supplement a carbon tax scheme with a border adjustment mechanism that would essentially apply a tariff to carbon-intensive, imported goods. Such a mechanism was included in H.R. 2454 (111th Congress). Under that proposal, EPA would have required importers of energy-intensive products from countries with insufficient carbon policies to submit a prescribed amount of "international reserve allowances," for their products to gain entry into the United States. Implementation of this approach would present substantial administrative challenges, particularly in terms of data from other nations. In addition, either the revenue distribution or border adjustment approach would likely raise concerns of trade complications.

CONCLUSION

Carbon taxes, or fees on emissions of some or all GHG emissions, have been proposed for many years by economists and some Members of Congress. A new carbon price would help reduce GHG emissions contributing to climate change and ocean acidification, and tax revenues could support a range of policy objectives, including deficit reduction.

Carbon tax revenues would depend strongly on the scope and rate of the tax and multiple market factors, which instill uncertainty in revenue projections. A \$20/mtCO_2 carbon tax on U.S. CO_2 emissions would generate approximately \$90 billion in its first year. If applied toward deficit reduction, carbon tax revenue (of this magnitude) could have some impact on projected budget deficits, but impacts vary considerably depending on which budgetary baseline is assumed. Regardless, if policymakers established a carbon tax, they would likely face pressure from multiple stakeholders seeking a portion of the carbon tax revenues. Households would be expected to bear a large portion of burden imposed by a carbon tax. Lower-income households, in particular, would face a disproportionate impact if revenues were not recycled back to them in some fashion. In addition, specific industries may experience disproportionate impacts. Carbon tax revenues that are used to offset the burden imposed on various sectors or specific population groups would not be available to support other objectives, like deficit reduction.

APPENDIX. CARBON TAX AND CARBON PRICING PROPOSALS IN THE 111TH CONGRESS

In the 111th Congress, policymakers introduced at least 9 stand-alone, market-based proposals that sought to reduce GHG emissions.[98] These bills included carbon tax, cap-and-trade, and hybrid approaches. Of the bills that specified a distribution formula for carbon tax revenue or emission allowance value,[99] only three (and one draft bill) would have allotted allowance value or tax revenue that would explicitly support deficit reduction (listed in order of proposed date):

- H.R. 2454 (Waxman-Markey), introduced May 15, 2009: 0.2% of emission allowance value in 2016-2026, zero in subsequent years.
- S. 1733 (Kerry-Boxer), introduced September 30, 2009: 10.3% of emission allowance value in 2016, increasing to 30% in 2030.
- S. 2877 (Cantwell), introduced December 11, 2009: 25% of emission allowance value (subject to the appropriations process) to a fund, which could be used to support a myriad of policy objectives, including deficit reduction.
- Kerry-Lieberman draft legislation, released May 12, 2010: 6.8% of emission allowance value in 2016 and 2030.

End Notes

[1] Congressional Budget Office, *An Update to the Budget and Economic Outlook: Fiscal Years 2012 to 2022*, Washington, DC, August 2012, http://www.cbo.gov/publication/43539.

[2] For background on budget baselines, see CRS Report R42362, *The Federal Budget: Issues for FY2013 and Beyond*, by Mindy R. Levit and CRS Report R41778, *Reducing the Budget Deficit: Policy Issues*, by Marc Labonte.

[3] See CRS Report RL33657, *Running Deficits: Positives and Pitfalls*, by D. Andrew Austin.

[4] A sampling of recent news and academic articles includes Mark Golden and Mark Shwartz, "Stanford's George Shultz on Energy: It's Personal," Stanford University, July 12, 2012; Jon Greenberg, "Laffer Carbon Tax: A Carbon Tax With a Twist to Please GOP—Maybe," *New Hampshire News*, December 14, 2011; Robert Inglis, "Fixing Market Distortions: A Free-Market Solution for Energy and Climate," University of Chicago Booth School of Business, April 11, 2012; Gilbert Metcalf, "Designing a Carbon Tax to Reduce U.S. Greenhouse Gas Emissions" *Review of Environmental Economics and Policy 3*, no. 1, pp. 63-83, September 2008; Brad Plumer, "Romney's One Big Idea on Climate—and Why He's Unlikely to Pursue It," *The Washington Post*, March 8, 2012; Amy Wolf, "Economist Arthur Laffer Proposes Taxing Pollution Instead of Income," Vanderbilt University, February 20, 2012.

[5] See the Save Our Climate Act of 2011 (H.R. 3242) and the Managed Carbon Price Act of 2012 (H.R. 6338).

[6] For a review of carbon taxes levied in other countries, see Jenny Sumner, et al., *Carbon Taxes: A Review of Experiences and Policy Design Considerations*, National Renewable Energy Laboratory, NREL/TP-6A2-47312, December 2009.

[7] See, e.g., House Committee on the Budget Chairman Paul Ryan's "Path to Prosperity" report, released to accompany the FY2012 Budget Resolution (H.Con.Res. 34), which suggests eliminating unspecified tax expenditures to allow for reduced marginal tax rates. This report also states that eliminating tax expenditures would not be for the purpose of generating

additional tax revenues. President Obama also expressed support for a revenue-neutral corporate tax reform as part of his 2012 State of the Union Address. Text of this address is available online at http://www.whitehouse.gov/ the-press-office/2012/01/24/remarks-president-state-union-address.

[8] CRS Report R42435, *The Challenge of Individual Income Tax Reform: An Economic Analysis of Tax Base Broadening*, by Jane G. Gravelle and Thomas L. Hungerford.

[9] Both the Fiscal Commission and Debt Reduction Task Force plans would increase the wage base upon which payroll taxes are applied, thus increasing payroll tax revenues.

[10] The Fiscal Commission's deficit reduction proposal included a $0.15 per gallon increase in the federal motor fuel excise tax.

[11] See CRS Report R41602, *Should the United States Levy a Value-Added Tax for Deficit Reduction?*, by James M. Bickley and CRS Report R41708, *Value-Added Tax (VAT) as a Revenue Option: A Primer*, by James M. Bickley.

[12] Carbon dioxide (CO_2) accounted for approximately 84% of U.S. GHG emissions in 2010. Approximately 94% of the CO_2 emissions resulted from fossil fuel combustion activities. See U.S. Environmental Protection Agency, *Inventory of U.S. Greenhouse Gas Emissions and Sinks*, 1990-2010, April 2012, at http://epa.gov/climatechange/ghgemissions/ us inventoryreport.html.

[13] For more information on climate change science, see CRS Report RL34266, *Climate Change: Science Highlights*, or CRS Report RL33849, *Climate Change: Science and Policy Implications*, both by Jane A. Leggett.

[14] See, for example, in 2007, the InterAcademy Council and the International Council of Academies of Engineering and Technological Sciences—both representing academies of sciences from dozens of nations—issued statements concurring that warming of the climate over the past 50 years is likely to have been caused by increased concentrations of GHG emissions in the atmosphere, with that increase having been caused by human-related GHG emissions. The views of a relatively small circle of scientists have been widely publicized as opposing mainstream conclusions; however, for these individuals the disagreement lies not in whether GHG increases have been caused by human activities, but whether these exert a "dangerous" influence on climate. See, for example, Robert M. Carter, "Knock, Knock: Where Is the Evidence for Dangerous Human-Caused Global Warming?" *Economic Analysis & Policy*, September 2008; S. Fred Singer, "Human Contribution to Climate Change Remains Questionable." *EOS Transactions* 80, April 20, 1999, pp. 183–187.

[15] Regulatory approaches can also be used to address market failures. Regulation-based emissions controls that reduce emissions would also be expected to increase the price of fossil fuels. The higher price faced by consumers may be closer to the true cost associated with consumption of emissions-producing fossil fuels.

[16] For a discussion see CRS Report R40242, *Carbon Tax and Greenhouse Gas Control: Options and Considerations for Congress*, by Jonathan L. Ramseur.

[17] Carbon tax proposals that apply only to CO_2 generally attach a price to a metric ton of CO_2 emissions (tCO_2). Non- CO_2 GHG emissions could be addressed by attaching a price to a metric ton of CO_2 emissions-equivalent (tCO_2e). This term of measure is used because GHGs vary by global warming potential (GWP). GWP is an index of how much a GHG may contribute to global warming over a period of time, typically 100 years. GWPs are used to compare gases to CO_2, which has a GWP of 1. For example, methane's GWP is 25, and thus a ton of methane is 25 times more potent a GHG than a ton of CO_2.

[18] A non-carbon example could be taxing nitrogen-based fertilizers based on their propensity to be emitted from agricultural soils as nitrous oxide (N2O), another GHG the emissions of which would be difficult to measure directly.

[19] There are additional GHGs, but they are smaller influences (though rapidly growing in some cases) and less measured. Also, this does not account for the emissions and removals of carbon dioxide from the atmosphere by land use, land use change, and forestry (LULUCF). For more information, see CRS Report RL34266, *Climate Change: Science Highlights*, by Jane A. Leggett.

[20] An upstream approach would apply a carbon tax to fossil fuels when they enter the U.S. economy, either at the mine, wellhead, or another practical "chokepoint" in the production chain, such as oil refineries.

[21] See 40 CFR Part 98.

[22] EPA, Fact Sheet: Mandatory Reporting of Greenhouse Gases (40 CFR Part 98), June 2011.

[23] The ability of a carbon tax to contribute to deficit reduction is analyzed in greater detail elsewhere in this report. See the section "Contribution to Deficit Reduction" below.

[24] Congressional Budget Office, *The 2012 Long-Term Budget Outlook*, June 2012.

[25] The SCC has been estimated for the purposes of allowing government agencies to consider social benefits from reduced carbon emissions in cost-benefit analysis. Details on the SCC for this purpose can be found in Interagency Working Group on the Social Cost of Carbon, United States Government, *Technical Support Document: Social Cost of Carbon for Regulatory Impact Analysis Under Executive Order 12866*, February 2010. SCC estimates are also provided in William D. Nordhaus, *Estimates of the Social Cost of Carbon: Background and Results from the Rice-2011 Model*, National Bureau of Economic Research, Working Paper 17540, Cambridge, MA, October 2011.

[26] These costs may be estimated simultaneously, or separately, as is done in National Academy of Sciences, *Hidden Costs of Energy: Unpriced Consequences of Energy Production and Use*, 2010; and Michael Greenstone and Adam Looney, *A Strategy for America's Energy Future: Illuminating Energy's Full Costs*, The Hamilton Project, Strategy Paper, May 2011.

[27] A critical and controversial assumption here is the discount rate used to discount future dollars into a present value. The use of low discount rates, such as some analyzed in Nicholas Stern's 2007 *Review on the Economics of Climate Change*, imply that aggressive policy action should be taken today to address future costs associated with climate change. Higher discount rates, such as those favored by William Nordhaus, would recommend applying fewer of today's resources to addressing climate change in the future (see William D. Nordhaus, "A Review of the Stern Review on the Economics of Climate Change," *Journal of Economic Literature*, vol. 45, no. 3 (September 2007), pp. 686-702.) Higher discount rates mean that costs in the future are worth less today, and less investment is warranted in the present to address future costs of climate change impacts. On the other hand, a fairness issues is raised by some in that high discount rates place little value on costs borne by generations yet to be born, whose preferences are generally not reflected in current generations' discount rates. There is little agreement in the economics community on appropriate discount rates to value such inter-generational problems.

[28] For further details see CRS Report R40242, *Carbon Tax and Greenhouse Gas Control: Options and Considerations for Congress*, by Jonathan L. Ramseur. See also the section on economics-based policy approaches in CRS Report R41973, *Climate Change: Conceptual Approaches and Policy Tools*, by Jane A. Leggett.

[29] H.R. 2454 (the American Clean Energy and Security Act of 2009, often called the "Waxman-Markey" bill) passed the House on June 26, 2009.

[30] For more information, see CRS Report R40809, *Climate Change: Costs and Benefits of the Cap-and-Trade Provisions of H.R. 2454*, by Larry Parker and Brent D. Yacobucci.

[31] To "hedge" is to insure oneself against losses. Investors typically face uncertainty and may use "hedging" as a strategy to reduce losses if/when adverse conditions occur. Investors may take actions that would offset the risk of another investment. Many analysts have recommended that policymakers hedge against future climate change risks by making investments, perhaps at a lower rate of return than other options, that would make sense whether or not future climate change is seriously adverse. For example, investing in low-emitting technologies may be one kind of hedge against climate change; stimulating such investments through a carbon tax may also be considered a hedge by many.

[32] For explanation of how carbon emissions may acidify oceans and the potential effects, see CRS Report R40143, *Ocean Acidification*, by Eugene H. Buck and Peter Folger.

[33] Most legislative attention to reduce GHG emissions, however, has been given to emissions cap-and-trade schemes and tax incentives. See for example, CRS Report R40242, *Carbon Tax and Greenhouse Gas Control: Options and Considerations for Congress*, by Jonathan L. Ramseur; CRS Report R40556, *Market-Based Greenhouse Gas Control: Selected Proposals in the 111th Congress*, by Brent D. Yacobucci, and Jonathan L. Ramseur. For other policy instruments, see CRS Report R41973, *Climate Change: Conceptual Approaches and Policy Tools*, by Jane A. Leggett, and CRS Report R41769, *Energy Tax Policy: Issues in the 112th Congress*, by Molly F. Sherlock and Margot L. Crandall-Hollick.

[34] Non-carbon GHGs could be subject to the tax based on their contribution to global warming in relation to CO_2. Global warming potential (GWP) is an index of how much a GHG may contribute to the average annual increase in worldwide temperature integrated over a period of time, typically 100 years but sometimes as short as 20 years. GWPs are used to compare gases to carbon dioxide, which has a GWP of 1. For example, methane's GWP is 25, and is thus 25 times more potent a GHG than CO_2. The GWPs listed in this report are from: Intergovernmental Panel on Climate Change, *Climate Change 2007: The Physical Science Basis,* 2007, p. 212.

[35] The Congressional Budget Office defines user charges as fees or taxes that are based on benefits individuals or firms receive from the federal government or that in some way compensate for costs they might impose on society or its resources. See Congressional Budget Office, *The Growth of Federal User Charges*, August 1993. See also GAO, *Federal User Fees: A Design Guide*, GAO-08-386SP, May 2008.

[36] See, for example, the evaluation criteria identified in American Institute of CPAs. "AICPA Tax Policy and Tax Reform Materials: No.1 Guiding Principles of Good Tax Policy: A Framework for Evaluating Tax Proposals", 2001, available at http://www.aicpa.org /INTERESTAREAS/TAX/RESOURCES/TAXLEGISLATIONPOLICY/Pages/ TaxReform.aspx.

[37] See this categorization in, for example, Joseph T. Sneed, "The Criteria of Federal Income Tax Policy," *Stanford Law Review* 17, no. 4 (April 1, 1965) pp. 567-613; B.G. Peters, *The Politics of Taxation: A Comparative* Perspective, 1991, at http://www.tau.ac.il/law/cegla3 /tax2/9%20The%20Politics%20of%20Taxation%20-%20Peters.pdf.

[38] For example, the Taxpayer Protection Pledge promoted by the Americans for Tax Reform interest group; Information at http://www.atr.org/taxpayer-protection-pledge and http://www.atr.org/about-grover.

[39] For data on federal excise tax collections, see Internal Revenue Service (IRS), Statistics of Income (SOI) Historical Table 20, http://www.irs.gov/taxstats/article/0,,id=175900,00.html.

[40] In this proposal, a ton is a short ton, or 2,000 pounds. Most of the more recent proposals and models use metric tons (or tonnes), which are 2,240 pounds.

[41] The Committee for a Responsible Federal Budget's (CRFB) online comparison tool is a good starting point for comparing various debt and deficit reduction proposals. This tool is available at http://crfb.org/compare/.

[42] Bipartisan Policy Center, *Restoring America's Future: Reviving the Economy, Cutting Spending and Debt, and Creating a Simple, Pro-Growth Tax System*, November 2010, http://bipartisanpolicy.org/projects/debt-initiative/about.

[43] Karen Palmer et al., *The Variability of Potential Revenue from a Tax on Carbon*, Resources for the Future, May 2012.

[44] Achieving economic efficiency means using limited resources such that the production of goods and services is maximized at the lowest possible cost.

[45] Though some people contend that GHG emissions may not pose adverse risks, this is the view of a minority of scientific and economic experts in the field of climate change, and the suggestion of no adverse risk is not supported by empirical evidence to date.

[46] This assumes that a new tax on carbon would ultimately apply to a broader base, and not simply compound existing distortions to labor and capital that exist in the current tax system. A stand-alone carbon tax could compound existing tax-induced distortions in labor markets, for example. Thus, a carbon tax could have efficiency costs to the extent that a carbon tax further reduces labor supply.

[47] For an extensive theoretical exposition of conditions under which a carbon tax can enhance economic efficiency, see A. Lans Bovenberg and Lawrence H. Goulder, "Environmental Taxation and Regulation," in *Handbook of Public Economics*, ed. Alan J. Auerback and Martin Feldstein, vol. 3 (Elsevier, 2002), pp. 1471-1545.

[48] A head tax, or a lump-sum tax levied on individuals regardless of income or wealth, is considered a non-distortionary tax since individuals cannot avoid the tax by changing their behavior. Since most taxes can be avoided with changes in behavior, non-distortionary taxes often do not exist in practice.

[49] In economics parlance, a small response of consumer demand may be termed "price inelastic."

[50] See, e.g., Gilbert Metcalf, *A Green Employment Tax Swap: Using a Carbon Tax to Finance Payroll Tax Relief*, 2007; Nathaniel Keohane and Sheila Olmstead, *Markets and the Environment* (Island Press, 2007); Ian Parry "Fiscal Interactions and the Case for Carbon Taxes over Grandfathered Carbon Permits," in *Climate Change Policy*, Dieter Helm, editor, (Oxford University Press, 2005).

[51] A theoretical exposition illustrating how marginal distortionary effects rise with the rate of taxation can be found in Jonathan Gruber, *Public Finance and Public Policy* (New York: Worth Publishers, 2007), pp. 584-585.

[52] Imposing regulations can also correct for market failures.

[53] A related discussion appears in the "Rate of Taxation" section above.

[54] For more information on the challenges of estimating the environmental and health costs of GHG emissions, see CRS Report R41973, *Climate Change: Conceptual Approaches and Policy Tools*, by Jane A. Leggett.

[55] This is in contrast to the statutory incidence, where the statutory incidence falls on the person responsible for remitting the tax to the tax authorities.

[56] See CRS Report R41641, *Reducing the Budget Deficit: Tax Policy Options*, by Molly F. Sherlock.

[57] See CRS Report R40841, *Assisting Households with the Costs of a Cap-and-Trade Program: Options and Considerations for Congress*, by Jonathan L. Ramseur and Libby Perl.

[58] See Stephen Moore, "Federal Budget Issue: Do We Need an Energy Tax?" National Center for Policy Analysis, June 1993, at http://www.ncpa.org/pub/bg127?pg=4, as an example of expressed concerns.

[59] During the debates of the 1993 Btu tax proposal, exemptions allowed to various interests may have contributed to a perception of its unfairness. For one view of the politics of the Btu tax debate and demise, see Dawn Erlandson, "The Btu Tax Experience: What Happened and Why It Happened," *Pace Environmental Law Review* 12, no. 1, 1994.

[60] C. Eugene Steuerle, "And Equal (Tax) Justice for All?" in *Tax Justice: The Ongoing Debate,* edited by Joseph J. Thorndike and Dennis J. Ventry (Urban Institute Press, 2002).

[61] Discussions of individual equity and benefit taxation arise mostly in discussions of public finance for education, Social Security, and other kinds of programs that offer an explicit benefit. See, for example, American Academy of Actuaries, "Social Security Reform: Changes to the Benefit Formula and Taxation of Benefits" at http://www.actuary.org /pdf/socialsecurity/benefit_05.pdf; or Kent E. Portney, "Individual Equity And School Finance: Implications For Taxation And State Aid." *Journal of Education Finance* 2, no. 2, 1976.

[62] A cap-and-trade system is arguably is more administratively complex than a tax. However, both systems, if they apply to all GHG and all large sources of emissions, involve many components. Moreover, as policymakers include more flexible design elements—primarily to improve efficiency and control price volatility—a cap-and-trade program would increase in complexity. It would be more costly to administer; more open to exploitation (for example, by traders of derivative financial instruments); potentially less transparent; and harder for regulators to ensure compliance.

[63] EPA's Greenhouse Gas Reporting Program in 40 CFR Part 98.

[64] See CRS Report R42392, *Aviation and the European Union's Emission Trading Scheme*, by Jane A. Leggett, Bart Elias, and Daniel T. Shedd.

[65] The CBO takes this into account when estimating net revenues from a carbon tax.

[66] See CRS Report R40936, *An Overview of Greenhouse Gas (GHG) Control Policies in Various Countries*, by Jane A. Leggett et al.

[67] See CRS Report RL34488, *Regulating a Carbon Market: Issues Raised By the European Carbon and U.S. Sulfur Dioxide Allowance Markets*, by Mark Jickling and Larry Parker.

68 The tax code already contains provisions allowing tax credits for carbon sequestration (see Internal Revenue Code §45Q).

[69] See CRS Report RL34436, *The Role of Offsets in a Greenhouse Gas Emissions Cap-and-Trade Program: Potential Benefits and Concerns*, by Jonathan L. Ramseur.

[70] See, for example, Dawn Erlandson, "The Btu Tax Experience: What Happened and Why It Happened" *Pace Environmental Law Review* 12, no. 1 (Fall 1994); Lisa Lerer, "Is Cap and Trade Dems' Next 'BTU'?" *Politico*, July 13, 2009; Joseph E. Aldy and Robert N. Stavins, "Using the Market to Address Climate Change: Insights from Theory and Experience," National Bureau of Economic Research Working Paper Series No. 17488 (2011).

[71] The 10-year budget deficit covers fiscal years 2013 through 2022.

[72] This scenario assumes that (1) most expiring tax provisions are extended and the Alternative Minimum Tax (AMT) is adjusted for inflation, (2) Medicare's payment rates for physicians' are held constant at current levels, and (3) that the automatic spending reductions required by the Budget Control Act (BCA) do not occur.

[73] One way to interpret this efficiency objective would be to maximize growth of GDP, though other measures of broader well-being may be an alternative. However, even using a narrow GDP measure would involve decisions about whether economic efficiency pertains to the

United States only or international efficiency, or to current taxpayers versus future generations.

[74] For more information, see CRS Report R40242, *Carbon Tax and Greenhouse Gas Control: Options and Considerations for Congress*, by Jonathan L. Ramseur and Larry Parker; and CRS Report R40841, *Assisting Households with the Costs of a Cap-and-Trade Program: Options and Considerations for Congress*, by Jonathan L. Ramseur and Libby Perl.

[75] This issue received considerable attention during the debate over H.R. 2454 ("Waxman-Markey") in the 111th Congress. H.R. 2454, which passed the House on June 26, 2009, would have, among other things, established a price on GHG emissions through a cap-and-trade system.

[76] National Commission on Energy Policy, *Allocating Allowances in a Greenhouse Gas Trading System*, 2007.

[77] As mentioned above, this is the expected outcome because lower-income households generally spend a higher percentage of their income on energy-related goods and services than do higher-income households.

[78] This study assumed a carbon tax of $15 per ton of CO_2.

[79] For more information on these different mechanisms, see CRS Report R40841, *Assisting Households with the Costs of a Cap-and-Trade Program: Options and Considerations for Congress*, by Jonathan L. Ramseur and Libby Perl.

[80] The per-capita, lump-sum rebate amount assumed in this study was $274.

[81] The credit would cover the first $560 in payroll taxes (or first $3,660 of wages per covered worker, using 2003 data).

[82] Under this option, the maximum credit amount would be $420.

[83] Studies that use this approach include Sebastian Rausch, et al., "Distributional Impacts of Carbon Pricing: A General Equilibrium Approach with Micro-Data For Households," *Energy Economics*, vol. 33, no. 1 (March 2011) pp. S20-S33 and Sebastian Rausch, et al., "Distributional Impacts of a U.S. Greenhouse Gas Policy: A General Equilibrium Analysis of Carbon Pricing," in *U.S. Energy Tax Policy*, ed. Gilbert E. Metcalf (Cambridge University Press, 2011), pp. 52-107.

[84] Sebastian Rausch, et al., "Distributional Impacts of a U.S. Greenhouse Gas Policy: A General Equilibrium Analysis of Carbon Pricing," in *U.S. Energy Tax Policy*, ed. Gilbert E. Metcalf (Cambridge University Press, 2011), pp. 52-107.

[85] Although often used as an economy-wide measurement, GDP is an imperfect measure of the economy or society's well-being. A classic example is that an epidemic may increase measured economic activity because of higher expenditures on health care, while reducing well-being. See also Organisation for Economic Co-operation and Development. How's Life?: Measuring Well-Being. Paris, 2011; Mack Ott, "Limitation of NIA as a Gauge of Welfare" in *The Concise Encyclopedia of Economics: National Income Accounts*, Library of Economics and Liberty, at http://econlib.org/library/Enc/NationalIncomeAccounts.html.

[86] For example, a reduction in GHG emissions from certain sectors may also entail a reduction in hazardous air pollutants, which could provide health-related benefits.

[87] Ian Parry and Robertson C. Williams III, *Moving U.S. Climate Policy Forward: Are Carbon Taxes the Only Good Alternative?* Resources for the Future, 2011; Warwick McKibben et al., *The Potential Role of a Carbon Tax in U.S. Fiscal Reform*, The Brookings Institution, 2012.

[88] For further information, see Ian Parry, "Fiscal Interactions and the Case for Carbon Taxes over Grandfathered Carbon Permits," in *Climate Change Policy* (Dieter Helm, editor), 2005; and

Lawrence H. Goulder, "Environmental Taxation and the Double Dividend: A Reader's Guide," *International Tax and Public Finance*, vol. 2, no. 2, 1995, pp. 157-183.

[89] Ian Parry and Robertson C. Williams III, *Moving U.S. Climate Policy Forward: Are Carbon Taxes the Only Good Alternative?* Resources for the Future, 2011.

[90] For example, the following reports found that a price or tax on carbon would have a small, but negative, impact on GDP growth: U.S. Energy Information Administration, *Energy Market and Economic Impacts of H.R. 2454, the American Clean Energy and Security Act of 2009*, Report #: SR-OIAF/2009-05, August 4, 2009, http://www.eia.gov/ oiaf/servicerpt /hr2454/background.html and Congressional Budget Office, *The Economic Effects of Legislation to Reduce Greenhouse-Gas Emissions*, Washington, DC, September 2009, http://www.cbo.gov/sites/default/files/cbofiles/ ftpdocs/105xx/doc10573/09-17-greenhouse-gas.pdf.

[91] This is referred to as the "tax-interaction effect" in economic literature. See, e.g., Ian Parry, "Fiscal Interactions and the Case for Carbon Taxes over Grandfathered Carbon Permits," in *Climate Change Policy* (Dieter Helm, editor), 2005.

[92] See, e.g., Ian Parry and Robertson C. Williams III, *Moving U.S. Climate Policy Forward: Are Carbon Taxes the Only Good Alternative?* Resources for the Future, 2011; Warwick McKibben et al., *The Potential Role of a Carbon Tax in U.S. Fiscal Reform*, The Brookings Institution, 2012.

[93] Note that this net economy-wide gain does not include benefits achieved by reducing GHG emissions.

[94] This study assumed a tax of $15 per ton of CO_2, with the tax rate rising at 4% above inflation each year through 2050. See Warwick McKibben et al., *The Potential Role of a Carbon Tax in U.S. Fiscal Reform*, The Brookings Institution, 2012

[95] For an analysis of the distribution of federal taxes, including corporate taxes, see Congressional Budget Office, *The Distribution of Household Income and Federal Taxes, 2008 and 2009*, Washington, DC, July 2012, http://www.cbo.gov/sites /default/files /cbofiles/attachments/43373-06-11-HouseholdIncomeandFedTaxes.pdf. The corporate tax is progressive, with 87.4% of the corporate tax burden attributable to the top two income quintiles. Since carbon taxes are regressive, and corporate taxes are progressive, using carbon tax revenues to reduce corporate taxes would tend to reduce the progressivity of the overall tax system.

[96] During debate over H.R. 2454 (in the 111th Congress), the Energy-Intensive Manufacturers' Working Group on Greenhouse Gas Regulation provided detailed testimony on the energy intensity and trade intensity of the U.S. manufacturing sector. Testimony of John McMackin for the Energy Intensive Manufacturers' Working Group on Greenhouse Gas Regulation before the House Committee on Ways and Means (March 24, 2009).

[97] For further details, see CRS Report R40914, *Climate Change: EU and Proposed U.S. Approaches to Carbon Leakage and WTO Implications.*

[98] See CRS Report R40556, *Market-Based Greenhouse Gas Control: Selected Proposals in the 111th Congress*, by Brent D. Yacobucci, and Jonathan L. Ramseur.

[99] Allowance value allocation includes both the use of auction revenues and distribution of allowances to entities (e.g., industry or non-covered parties) at no cost.

In: Carbon Taxes
Editor: Neil Viveiros
ISBN: 978-1-62618-148-9

Chapter 2

CARBON TAX AND GREENHOUSE GAS CONTROL: OPTIONS AND CONSIDERATIONS FOR CONGRESS*

Jonathan L. Ramseur and Larry Parker

SUMMARY

Market-based mechanisms that limit greenhouse gas (GHG) emissions can be divided into two types: quantity control (e.g., cap-and-trade) and price control (e.g., carbon tax or fee). To some extent, a carbon tax and a cap-and-trade program would produce similar effects: Both are estimated to increase the price of fossil fuels, which would ultimately be borne by consumers, particularly households. Although there are multiple tools available to policymakers that could control GHG emissions—including existing statutory authorities—this report focuses on a carbon tax approach and how it compares to its more frequently discussed counterpart: cap-and-trade.

If policymakers had perfect information regarding the market, either a price (carbon tax) or quantity control (cap-and-trade system) instrument could be designed to achieve the same outcome. Because this market ideal does not exist, preference for a carbon tax or a cap-and-trade program ultimately depends on which variable one wants to control—emissions or costs. Although

* *This is an edited, reformatted and augmented version of a Congressional Research Service publication, R40242, dated March 10, 2009.

there are several design mechanisms that could blur the distinction, the gap between price control and quantity control can never be completely overcome.

A carbon tax has several potential advantages. With a fixed price ceiling on emissions (or their inputs—e.g., fossil fuels), a tax approach would not cause additional volatility in energy prices. A set price would provide industry with better information to guide investment decisions: e.g., efficiency improvements, equipment upgrades. Economists often highlight a relative economic efficiency advantage of a carbon tax, but this potential advantage rests on assumptions—about the expected costs and benefits of climate change mitigation—that are uncertain and controversial. Some contend that a carbon tax may provide implementation advantages: greater transparency, reduced administrative burden, and relative ease of modification.

The primary disadvantage of a carbon tax is that it would yield uncertain emission control. Some argue that the potential for irreversible climate change impacts necessitates the emissions certainty that is only available with a quantity-based instrument (e.g., cap-and-trade). Although it may present implementation challenges, policymakers could devise a tax program that allows some short-term emission fluctuations, while progressing toward a long-term emission reduction objective. Proponents argue that short-term emission fluctuations would be preferable to the price volatility that might be expected with a cap-and-trade system.

Although a carbon tax could possibly face more political obstacles than a cap-and-trade program, some of these obstacles may be based on misunderstandings of the differences between the two approaches or on assumptions that the tax would be set too low to be effective. Carbon tax proponents could possibly address these issues to some degree, but there remains considerable political momentum for a cap-and-trade program.

INTRODUCTION

A variety of efforts that seek to reduce greenhouse gas emissions (GHG)[1] are currently underway or being developed on the international, national, and sub-national level (e.g., individual state actions or regional partnerships). One option (of many, see text box below—"Other Policy Options for Addressing GHG Emissions") for controlling GHG emissions is to apply a tax or fee on GHG emissions or the inputs that create them. This type of approach is commonly called (and referred to in this report as) a carbon tax,[2] whether it would apply to CO_2 emissions alone or to multiple GHGs, including some that may have no molecular carbon.[3] This report does not provide a comprehensive comparison and analysis of the multiple policy tools available to Congress that would address climate change. Instead, this report focuses on the policy considerations of using of a carbon tax to control GHG emissions.

Governments may impose taxes for a variety of purposes. The primary reason that governments impose taxes is to raise revenue to fund various objectives or services: e.g., national defense, public education, social security, etc. Generally, governments raise these revenue streams by placing a tax on activities that are recognized as desirable ("economic goods") such as income, employment, and investment. While this tax placement ensures a relatively steady flow of revenue (often the primary objective of the tax), economists generally describe such taxes as distortionary, because the taxes discourage the "good" activity. For example, many economists have argued that payroll and income taxes discourage employment and investment.[4] If these taxes were reduced, the incentives to increase labor and investment would be greater.

Economists maintain that levying a charge on pollution (sometimes referred to as a Pigouvian tax)[5] would be an efficient way to correct an inherent failure in a particular market. A basic economics principle is that market prices may not reflect the social cost of resource use (e.g., fossil fuel combustion) when economic activities result in pollution (e.g., CO_2 emissions). If social costs are not included, the market price of the resources will not reflect their true costs. For example, in terms of climate change policy, the price of using fossil fuels, particularly coal, does not reflect the costs—i.e., climate change-related damages—associated with CO_2 emissions.

In economics parlance, the social cost not reflected in the market price is called an "externality." A pollution tax would internalize the external costs by making the party who profits from the polluting activity include the external costs in the price of the good or service. Policymakers could place a pollution tax on GHG emissions or the inputs that create them. By attaching a price to GHG emissions, a carbon tax would stimulate GHG emission reduction. If the tax were placed on emissions, entities directly subject to the tax, such as power plants, would have an incentive to take actions—e.g., energy efficiency improvements or equipment upgrades—to lower tax payments. If the tax were placed on emission inputs—e.g., fossil fuels—the price of carbon-intensive energy sources, primarily coal, would increase relative to low-carbon fuels (**Table A-1** of this report—located in the **Appendix**—includes estimates of price increases to fossil fuels and motor gasoline based on different carbon tax rates). Energy consumers—e.g., power plants, industry, households, etc.—would be encouraged to (1) switch to less carbon-intensive fuels; (2) use less energy or use energy more efficiently; and (3) prefer products or services that are lower-priced by virtue of incorporating less emission tax. Each of these activities would reduce GHG emissions compared to a business-as-usual track.

Other Policy Options for Addressing GHG Emissions

For policymakers considering actions to address climate change, a variety of policy instruments is available. Although current attention has largely focused on market-based mechanisms, primarily cap-and-trade systems, non-market policy tools may be the most practical option to address some emission sources. Moreover, Congress may consider complementary approaches to market mechanisms to improve their effectiveness. In particular, many experts maintain that the GHG emission targets specified in recent legislation would require development and deployment of improved (low-carbon) technologies. To further this effort, some argue that Congress should address technology stimulation directly or as a supplement to the primary climate change mitigation policy. In addition, Congress has already addressed some specific sectors (cars and light trucks, and government buildings) through performance standards. These standards may be further strengthened, independent of climate change legislation.

Efforts are underway to address GHG emissions using the existing Clean Air Act (CAA) statute. The Environmental Protection Agency (EPA) has received at least eight petitions asking it to use its CAA authority to regulate GHG emissions from multiple categories of mobile sources. The agency faces lawsuits seeking GHG regulations (per CAA authority) from power plants, refineries, and other stationary sources. In 2007, the Supreme Court found (Massachusetts v. EPA) that GHG emissions are air pollutants under the CAA, and that EPA has the authority to promulgate GHG emission controls.

These expected behavioral changes mirror the activities that are forecast for a potential cap-and-trade program. Both a carbon tax and a cap-and-trade system would place a price on carbon. Both a carbon tax and cap-and-trade system are intended (and expected) to increase the price of coal, oil, and natural gas. Under either program, these price increases would ultimately be borne by energy consumers, both businesses and households. These price increases are integral to a market-based approach to GHG emission reduction, because they send more accurate information to purchasers about the full cost of their choices.

This report begins with an overview of the fundamental choices involved between a cost (tax) and a quantity (cap) control instrument. This includes a discussion of policy tools that could be employed to bridge the gap between a carbon tax and a cap-and-trade program. Following this overview, the report analyzes the potential advantages and disadvantages of a carbon tax. In many cases, carbon tax attributes are compared with those of a cap-and-trade program. The next section discusses implementation issues for a carbon tax, including where to apply the tax, at what level to set the tax, and options for distributing the tax revenues. The final section provides conclusions.

Cost of Quantity Control: An Overview

If policymakers choose to establish a market-based mechanism to control GHG emissions, a fundamental decision would be whether to use a price instrument, such as a carbon tax, or a quantity instrument, such as an emissions cap.[6]

Economic Theory vs. Uncertainty

In an economically efficient market with perfect information, either a price (carbon tax) or quantity control instrument (cap-and-trade system) could be designed to achieve the same outcome.[7] **Figure 1** illustrates this basic economic principle. The intersection of marginal costs[8] and marginal benefits[9] would provide the point (of economic efficiency) at which to set the price or quantity limit. At this point in the figure, the abatement costs[10] equal benefits received from abatement. Per economic theory, emission abatement above or below this point would not be economically efficient.[11] The dashed lines indicate the efficient price (tax) and quantity (cap) limits. This figure illustrates that (with perfect information) if either a tax or cap is selected, both the emission abatement level and cost of abatement would be identical. For instance, if a cap were chosen, covered sources (e.g., power plants) would abate emissions until they reach the cap of Q*, at which point the marginal cost of abatement would equal P*. If a tax were chosen, sources would abate emissions until the marginal cost of abatement reached the tax level (P*), at which point the emission abatement level equals Q*. In either case, total abatement is Q*, and the total cost of the program is the shaded area under the marginal abatement cost curve.

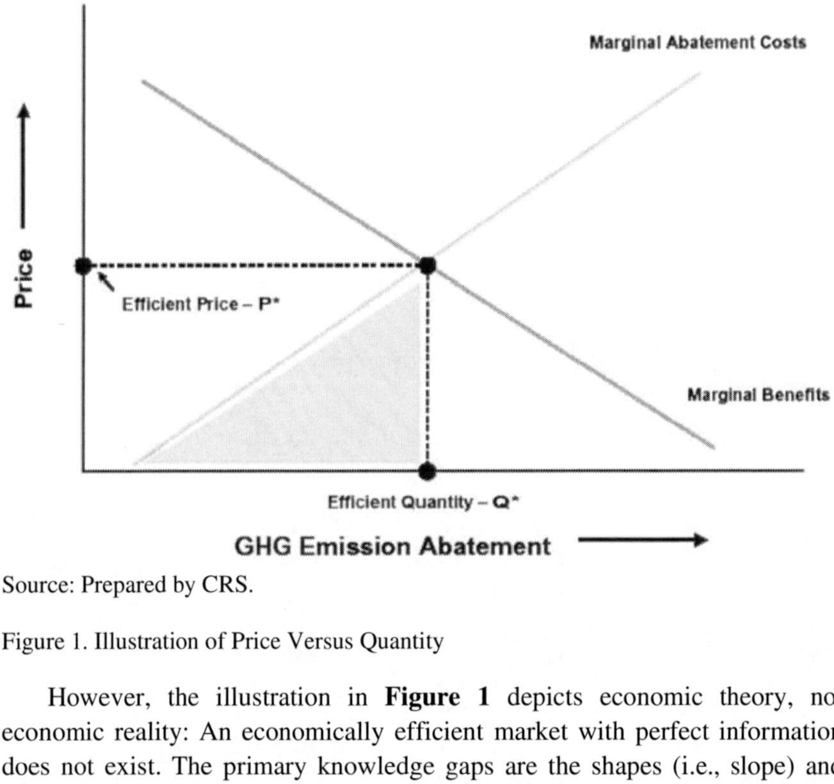

Source: Prepared by CRS.

Figure 1. Illustration of Price Versus Quantity

However, the illustration in **Figure 1** depicts economic theory, not economic reality: An economically efficient market with perfect information does not exist. The primary knowledge gaps are the shapes (i.e., slope) and levels (i.e., magnitude) of the marginal abatement cost and marginal benefit curves. These benefit and cost curves are influenced by multiple variables, including uncertain future emission baselines. Baseline estimates are related to projected population levels and economic growth,[12] factors that are also unknown and difficult to predict.[13] Another uncertain factor, which will have a substantial influence over mitigation costs, is the development and deployment of low-carbon technology.

Without accurate information regarding marginal costs and benefits, policymakers cannot know the efficient quantity or price from which to establish an emissions cap or carbon tax. Thus, in one sense, preference for a price (carbon tax) or a quantity limit (emissions cap) depends on one's preference for uncertainty—either uncertain emissions or uncertain program costs. A tax would set a ceiling on the marginal price of emissions (and thus the total cost of the program), but emissions would be uncertain. A cap would hold emissions to a set level, but the price of emission reductions (and thus program costs) would be uncertain.[14] However, as discussed below, the choice is not as simple as it may seem.

A Stark Choice or a Policy Continuum?

There are multiple design elements that can be included with a cap-and-trade program that blur the distinction between price and quantity control. Similarly, a carbon tax program could include flexible design mechanisms allowing policymakers to alter the tax rate, if they determine that emission reductions are not proceeding at a desirable pace. With these design elements available, Congress is presented with a policy continuum, rather than a stark policy dichotomy.

A Flexible Emissions Cap

Congress could enact a cap-and-trade program that includes mechanisms that provide flexibility in terms of controlling costs. **Figure 2** illustrates the options available. As the design elements in **Figure 2** (the "bridge pieces") are added to a cap-and-trade program, the program more closely resembles a carbon tax. These elements are briefly described below:

- Banking: Covered sources may save (bank) emission allowances[15] and submit them in future years. Banking reduces the absolute cost of compliance by making annual emissions caps flexible over time. In addition, banking reduces price volatility to some degree.[16]
- Borrowing: Covered sources may borrow (typically at a relatively high interest rate) emission allowances allotted to future years. The ability to borrow allowances from future years should help control price spikes.
- Market oversight: Congress could create a regulatory mechanism (e.g., an oversight board) to observe the allowance market and implement cost-relief measures if necessary. Obviously, the authority granted to the delegated entity would determine its ability to affect program costs.

- Safety valve: Generally triggered by prices in the allowance markets, safety valves may include (1) a set price alternative to making reductions or buying allowances at the market price, (2) a slowdown in tightening the emissions cap, and (3) lengthening of the time allowed for compliance. The first type would move the program closest to price control by setting a ceiling on allowance prices and limiting their volatility.
- Auctioning allowances: Auctions may be used to allocate some or all of the emission allowances. Like a carbon tax, auctions would raise revenues to support various objectives. In addition, auctions would support price discovery.[17]
- Reserve price: A reserve price is a price below which the seller refuses to part with the item for sale. A reserve price would all but guarantee a revenue stream in an emissions allowance auction, thus operating much like a minimum tax or price floor.

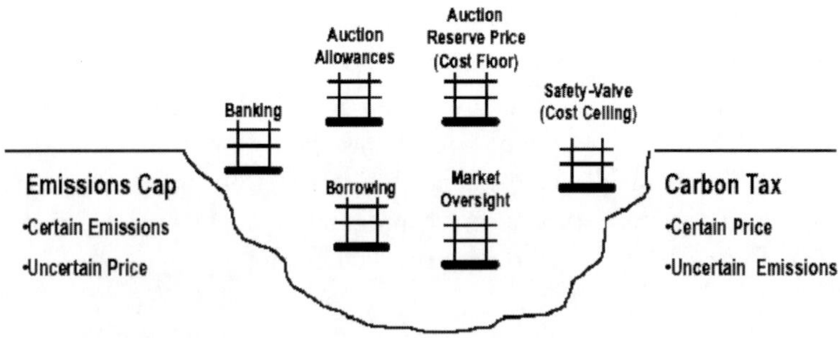

Source: Prepared by CRS.

Figure 2. Bridging the Gap between Price and Quantity Control

A Flexible Carbon Tax

Policymakers could establish a carbon tax that includes mechanisms for addressing long-term emission uncertainty. For example, emission sources would be required to submit emission data periodically, so that policymakers could assess the performance of the carbon tax. If Congress determined that emission reduction was inadequate, the tax rate could be adjusted to achieve a desired emissions level. The decision authority could be retained by Congress or delegated to an existing federal agency—e.g., Department of Treasury or Environmental Protection Agency—or another entity (e.g., a newly created independent board) in order to meet pre-determined emission reduction objectives.[18]

Considering the consequences of altering a carbon tax rate, Congress may be hesitant to delegate this authority. In addition, some may question whether the federal government is nimble enough to modify the carbon tax rate on a periodic basis, particularly considering the potential size of the tax (**Table 1** below). Moreover, a flexible carbon tax approach may sacrifice long-term price certainty, depending on the authority of the entity to adjust the tax rate. To address this concern, Congress may consider providing tax rate increase parameters for the delegated entity.

Limits of the Policy Continuum

Regardless of whether policymakers employ these additional design elements, the gap between price control and quantity control can never be completely overcome: The closest resemblance would be a cap-and-trade program that employs both a price safety-valve and an emission allowance auction with a reserve price. Even if all of the above design options are part of a cap-and-trade program, the price of emission allowances would still fluctuate to some degree. Likewise, a more flexible tax system would still yield short-term emission fluctuations.

The policy continuum demonstrates that a comparison between a carbon tax and an emissions cap is not a straightforward exercise. Design details are critical for an appropriate comparison. Depending upon program design, cap-and-trade programs can vary greatly in their ability to control price fluctuations and total program costs. Some of the advantages or disadvantages (depending on one's perspective) of either system can be addressed through program design. Indeed, recent federal cap-and-trade proposals have included one or more of the design elements listed above.

POTENTIAL ADVANTAGES OF A CARBON TAX

This report examines four potential advantages of a carbon tax approach to controlling GHG emissions. First, the report compares a carbon tax to cap-and-trade approach from an economics theory perspective, analyzing the assertion (of many economists) that a carbon tax would be more economically efficient than a cap-and-trade system. Second, it discusses the merits of price stability that a carbon tax would provide. The third discussion highlights the advantage of generating carbon tax revenues and examines various revenue applications and the trade-offs that would ensue. Finally, the report identifies potential implementation advantages, primarily practical considerations, of using a carbon tax approach.

Economic Efficiency

A primary argument supporting the use of a carbon tax is that in the presence of cost uncertainty—i.e., the marginal abatement cost curve (**Figure 1**)—a tax would potentially be more economically efficient than an emissions cap. However, this argument is grounded in assumptions that are the subject of considerable debate and controversy. This section outlines the basis for the argument and provides an analysis of the underlying assumptions.

Basis for the Argument

As **Figure 1** demonstrates, if the marginal cost and benefit curves are known to policymakers, either a tax or a cap instrument could be used to achieve the same result. As discussed above, both the costs and benefits of GHG emission reduction are uncertain. However, some economists argue that the *relative* slopes of the cost and benefit curves are better understood. This information—the slope of the marginal benefit curve in comparison to the slope of the marginal cost curve—is the foundation for an argument that supports a tax (price control) over an emissions cap (quantity control).

This argument was presented in Weitzman's landmark 1974 study "Prices vs. Quantities."[19] Weitzman's primary conclusion (explained in some detail below) was that price instruments (e.g., carbon tax) would be preferred when the marginal benefit curve is relatively flat; a quantity limit (e.g., emissions cap) would be preferred when the marginal cost curve is relatively flat.[20]

Intuitively, a relatively flat marginal benefits curve suggests that each additional unit of abatement provides approximately the same benefit. In contrast, a relatively steep marginal cost curve suggests that each additional unit (i.e., ton of CO_2) of emission abatement entails a greater cost than the previous unit.[21] Some economists argue that the slope of the marginal benefits curve is flat relative to the marginal cost curve.[22] This notion is the subject of intense debate, but it is grounded in the scientific processes (discussed below) through which GHG emissions generate climate change impacts.

Based on this (assumed) relationship between the expected slopes of costs and benefits, a carbon tax would be more efficient than an emissions cap. **Figure 3** and **Figure 4** below illustrate this concept. **Figure 3** provides a scenario in which the marginal benefits curve is flatter than the marginal costs curve. Without precise information regarding marginal costs and benefits, policymakers would base a tax or cap on *estimates* of the marginal cost and benefit curves. These are represented in **Figure 3** as the *expected* marginal costs and benefits of GHG emission abatement.[23] The intersection of the expected marginal abatement cost and benefit curves would be the estimated economically efficient point of regulation. From this efficient point, policymakers could apply either an emissions cap or carbon tax—at Q* or P*, respectively. Either option would produce the same outcome, if the *expected* marginal cost curve matches the *actual* marginal cost curve.[24]

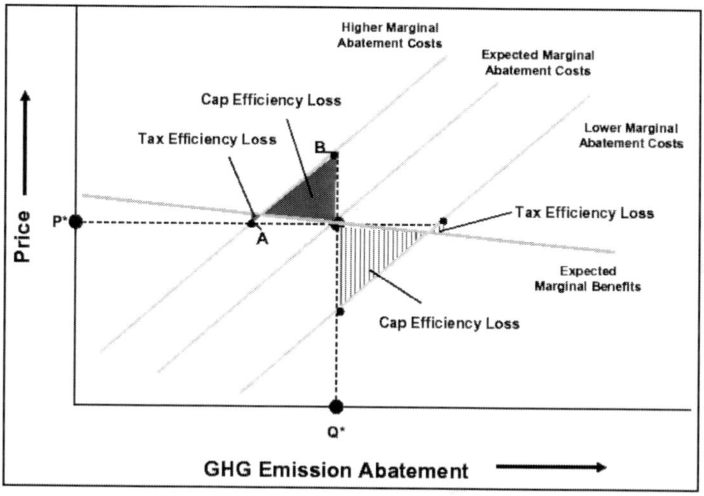

Source: Prepared by CRS.

Figure 3. Illustrative Scenario with a Relatively Flat Marginal Benefits Curve

However, the *actual* cost curve is unlikely to match the *expected* cost curve. Therefore, some level of economic inefficiency will exist with either a tax or an emissions cap. However, the policies are expected to yield different levels of inefficiencies. For example, consider a scenario (**Figure 3**) in which *actual* marginal abatement costs are higher than *expected.* If policymakers were to apply a carbon tax (P* in the figure) directly to emissions, emission sources would abate emissions until the marginal cost of abatement equals the carbon tax. This occurs at **Point A**, at which time covered sources would pay the tax for any additional emissions. On the other hand, if policymakers employed an emissions cap (Q*), covered sources would abate emissions until **Point B**, when the cap is met.

With a tax instrument in **Figure 3**, under higher-than-expected costs, covered sources would abate less than the optimal level—the intersection of benefits and *actual* costs; with an emissions cap, sources would abate more than the optimal level. However, **Figure 3** illustrates that the difference in efficiency losses—the shaded triangles—from the two policies is substantial.

Figure 3 also illustrates the scenario of lower-than-expected marginal abatement costs. With lower-than-expected marginal abatement costs, an emissions cap would lead to abatement that is less than optimal, whereas a tax would stimulate abatement beyond the optimal level. As with higher-than-expected costs, a tax instrument would yield less inefficiency than an emissions cap. The different efficiency losses are depicted by the striped triangles in **Figure 3**.

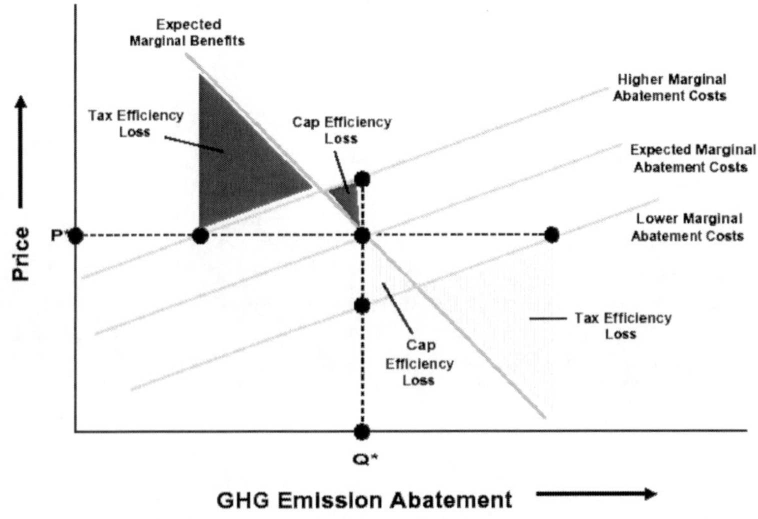

Source: Prepared by CRS.

Figure 4. Illustrative Scenario with a Relatively Steep Marginal Benefits Curve

Figure 4 illustrates the alternate scenario, in which the marginal benefits curve is relatively steep. In this situation, an emissions cap is the preferred instrument. Whether the marginal abatement costs are higher or lower than expected, the efficiency losses from a carbon tax are greater than those from an emissions cap.

Underpinnings of the Argument

The argument that a carbon tax would minimize economic inefficiency in the face of uncertain costs contains several assumptions. A primary assumption concerns the relative steepness/flatness of the cost and benefit curves. In addition, there are other underpinnings of the argument that may be questioned. These issues are discussed below.

Relative Flatness of Benefits Curve?

At the crux of the argument that a tax would maximize efficiency is the assumption that the slope of the marginal benefits curve is flat compared to the marginal costs curve (**Figure 3**). Many would assert this is the case, based on the process by which GHG emissions generate damages. In the context of climate change, the damages related to GHG emissions are due to their concentration (referred to as "stock") in the atmosphere, as opposed to their annual flow.[25] By comparison, more traditional air pollutants (e.g., sulfur dioxide, nitrogen oxide) impose damages through their annual emission flows.

With a "stock pollutant" like CO_2, the processes involved in adding or removing the gas from the atmosphere are measured, not in years, but in decades (or centuries for some gases). The current stock of CO_2 in the atmosphere is approximately 3,000 gigatons;[26] in 2004, the entire world emitted roughly 30 gigatons of CO_2. Because emission quantities would have only a minor effect on the accumulation of CO_2 in the atmosphere, the benefits from controlling the stock (or the damages from business-as-usual activities) are thought to occur slowly and steadily.[27] **Figure A-1** in the **Appendix** illustrates this concept.

However, the assertion that the benefits curve is flat assumes that there are no climatic "tipping points." Some scientists maintain that at a certain stock level—concentration of GHG emissions in the atmosphere—particularly drastic events might occur.[28] By definition, a tipping point implies that when it is reached, the course of climate change is abrupt and the marginal benefits curve becomes very steep. The climate changes may even be irreversible and catastrophic, at least for some populations and environments. Examples often cited include collapse of Greenland's ice sheet; and dieback of the Amazon rainforest.[29]

If such tipping points exist, they would dramatically alter the shape of the benefits curve, because abating emissions to avoid large adverse consequences would yield large benefits. As argued by the Stern Review: "over the long term, as the stock of GHGs grows, marginal damages are likely to rise and – as the stock reaches critical levels – marginal damages may rise sharply. In other words, the damage function is likely to be strongly convex."[30] However, the Stern Review also notes that: "to the extent that damages may relate to the *rate* of climate change, the relationship is more complex, but it remains true that the damage curve is likely to respond most to cumulative emissions over several years or even decades."[31] Depending on whether the tipping points are "several years" or "decades" away, an emissions cap may be preferred.

However, scientists and policymakers have very poor information regarding the GHG concentration levels (or rates of concentration changes) at which tipping points may occur. A 2003 economic study found that emissions caps would be "optimal only when there is a well-defined threshold with clearly catastrophic consequences on the other side."[32]

Relative Steepness of the Marginal Cost Curve?

The marginal cost curve is perhaps better understood (and less controversial) than benefits estimates. Analysis of recent cap-and-trade proposals introduced in Congress suggest that with the current portfolio of technologies the marginal cost curve does not appear flat.[33] Although the first units of emission abatement opportunities may be relatively inexpensive, the costs of abatement are likely to increase with each additional unit. However, if a carbon market spurs technology development, the marginal cost curve may latten. This could reverse the relationship between costs and benefits, and make a cap the more economically efficient policy.[34]

Short-Term Time Frame

In general, Weitzman's economic principle regarding the relative slopes of the marginal cost and benefit curves applies to a static or fixed time-frame. The principle is best suited to compare the costs and benefits of controlling a "flow pollutant" (such as SO_2), which would yield short-term damages. However, as noted above, CO_2 is a stock pollutant: its potential damages are related to total accumulation in the atmosphere; one year of emissions would likely have negligible effects. Thus, the static analysis is over-simplistic, because (by definition) it assumes that activities in different time periods are unrelated.[35] However, equipment upgrades or efficiency improvements made in one time period would affect the emissions generated in future years as well. Other time-related factors need to be considered, including the discounting of future benefits and the atmospheric decay of GHGs.

Several economic studies have addressed this issue of time, expanding Weitzman's basic principle within a dynamic framework (i.e., a multi-year time-period). In general, these studies concluded that Weitzman's basic principle carries over to a dynamic setting.[36]

Only Cost Uncertainty is Considered

Although marginal benefit information is also uncertain, the relative-slope principle generally depends only on the marginal cost curve. The rationale behind this is that when covered sources make decisions regarding emission abatement—e.g., reduce onsite emissions or pay the carbon tax—benefits of abatement are not part of their economic calculation. Economists point out that if covered sources did consider benefits, there would be no externality in the first place.

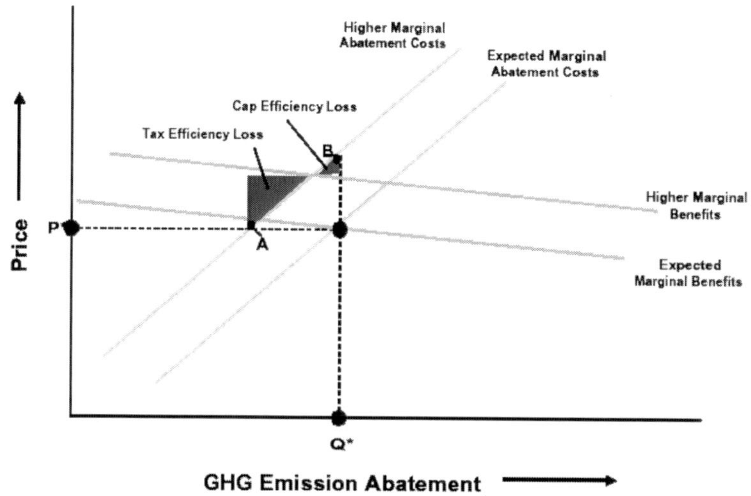

Source: Prepared by CRS.

Figure 5. Illustrative Scenario with Marginal Costs and Marginal Benefits That Are Higher Than Expected.

However, in some circumstances an emissions cap may produce a more economically efficient outcome. As mentioned (in a footnote) by Weitzman (1974) and re-examined by Stavins in 1996,[37] if the uncertainty of costs and benefits are positively correlated—e.g., both the marginal benefit and cost curves are actually higher than expected—quantity control provides a more economically efficient result. This scenario is illustrated in **Figure 5**. The advantage for quantity control over price control under these circumstances depends on the degree of correlation and the magnitude of uncertainty with respect to the cost and benefits curves – the greater the difference between anticipated and actual cost and benefits values, the greater the advantage for quantity control measures. Stavins points out that this reversal in preference for price over quantity control could occur with plausible values for the two variables; however, he argues that "the advantage of price over quantities mains unless the true benefits of carbon mitigation are many orders of magnitude greater than our best estimate."[38]

Modeled Efficiency Gains

Several economic studies have examined the efficiency gains of using a carbon tax versus an emissions cap.

> While the analysis is more complicated and involves more than simply the relative slopes of marginal abatement and damage curves, *the analyses consistently find that taxes dominate cap-and-trade systems for a broad range of parameter values* consistent with scientific understanding of the global warming problem [emphasis added].[39]

For example, Pizer found that taxes are more efficient than an emissions cap by a factor of five to one.[40] Although this result is from 1997, similar estimates were made in more recent studies that applied different parameters.[41] The magnitude of these efficiency gains were estimated to be in hundreds of millions to several billions of dollars per year.[42]

However, a 2008 Massachusetts Institute of Technology (MIT) report points out that the modeled efficiency gains may be smaller in practice.[43] As discussed above, policymakers can design cap-and-trade programs with flexible components, such as banking and borrowing, that could alleviate some of the inefficiency loss. But a 2008 Resources for the Future (RFF) study indicates that banking may not help "as much as proponents might suggest." The study found that banking provisions in a cap-and-trade program would improve efficiency by 20% over a cap-and-trade without banking.[44]

Economic Efficiency versus Precaution

Despite the economic efficiency argument in favor of a carbon tax, some argue that the downside risks of tipping points—i.e., irreversible climate changes—demand that economic efficiency take a backseat to precise control of GHG reductions, achievable only with a quantity-based emissions cap. This viewpoint could be described as an application of the precautionary principle: i.e., keeping one's options open in the face of uncertainty.[45] This approach argues that, while GHG emissions are held at an exact level, scientists and economists can continue to gather information regarding abatement costs and climate change benefits. As (presumably) better information is garnered, climate change policy can be amended as necessary.

Price Stability

Depending on the chosen GHG control instrument—carbon tax or cap-and-trade—either emissions or prices would fluctuate over short-term time periods (months to years). Proponents of a carbon tax often highlight the volatile price swings that have been observed in the U.S. sulfur dioxide (SO_2) emissions cap-and-trade program.[46] Between 2001 and 2006, the price of a SO_2 allowance has varied by a factor of 12.[47] The inclusion and substantial use of banking has not been sufficient to overcome other factors creating price volatility, such as uncertainty regarding proposed regulatory changes in the program (e.g., The Clean Air Interstate Rule).

Although the price volatility in the SO_2 market did not translate into electricity price spikes for consumers, CO_2 emission allowances could play a more integral role in the U.S. economy: Fossil fuels accounted for 85% of energy consumption in 2006.[48] Nordhaus states that strict quantity limits might have "major disruptive effects on energy markets and investment planning, as well as on the distribution of income across countries, inflation rates, energy prices, and import and export values."[49] Volatile emission allowance prices may discourage investment, because firms would have less reliable cost information upon which to plan for capital improvements.[50]

A GHG cap-and-trade program may or may not have similar price swings to those experienced with the SO_2 allowance market. The largest GHG cap-and-trade program currently operating is the European Union's Emission Trading System (EU ETS). During the four years of its existence, both Phase 1(2005-2007) and Phase 2 (2008-2012) EU ETS allowance prices have experienced significant volatility. Phase 1 allowance prices from the EU ETS are arguably not an appropriate comparison, because the fluctuations were largely related to program design issues.[51] Building on the experience gained during Phase 1, Phase 2 of the ETS began in 2008, and these emission allowance prices may provide a better example of the potential effects of market forces and the fluctuations that could occur under a U.S. cap-and-trade program.[52] **Figure 6** illustrates the market prices for Phase 2 allowances between 2005 and January 2009.

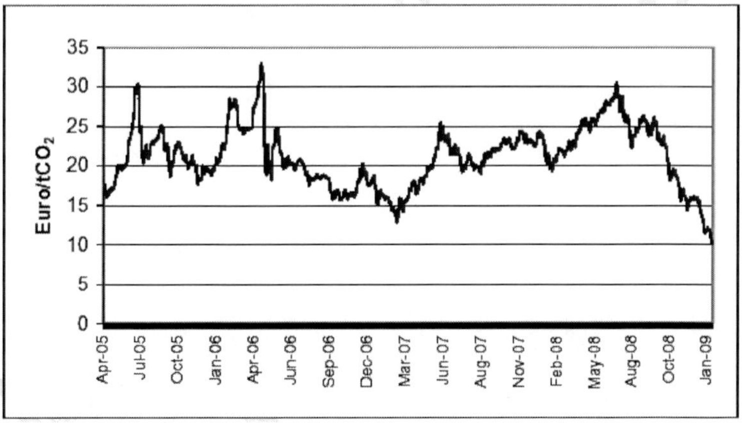

Source: Prepared by CRS with data from European Climate Exchange, at http://www.ecxeurope.com.

Figure 6. "Phase 2" Emission Allowance Prices in the European Union's Emission Trading System

Tax Revenue Applications

Carbon tax proponents often cite the generation of carbon tax revenues as a primary advantage of a carbon tax. However, policymakers could design a cap-and-trade program that could deliver similar, if not identical, results. For example, if policymakers establish a cap-and-trade program and distribute 100% of the allowances through an auction process, the auction proceeds would function as tax revenues.[53] Thus the strength of this advantage depends on the design of the cap-and-trade program to which the tax is compared.

From a public finance perspective, neither carbon tax nor auction revenues would be a reliable source of funding, because an effective carbon price (established through a tax or a cap) is expected to alter consumer behavior (e.g., demand for carbon-intensive goods). However, a carbon tax revenue stream would have a more predictable source than an allowance auction revenue stream, because the tax would be a known value, while the emission allowance price would fluctuate (as described above).

Different options for applying the tax revenue are discussed in the section "Implementation of a Carbon Tax."

Potential Implementation Advantages

Depending on the design, a carbon tax may offer several implementation advantages over a cap-and-trade program. These are discussed below.

Transparency

Arguably, the concept of a carbon tax is easier to explain and understand than a cap-and-trade program. A cap-and-trade system may have more moving parts, which may be challenging to follow, particularly the emission allowance allocation scheme. Moreover, as policymakers include more flexible design elements—primarily to improve efficiency and control price volatility—a cap-and-trade program would increase in complexity and potentially become less transparent. Some contend that the relative transparency of a carbon tax would help garner support for its enactment.[54] This argument may gain strength in the context of the recent financial climate, as many have grown more skeptical of seemingly complex financial structures.[55]

Although the concept of a carbon tax is arguably a simpler approach, many argue that the U.S. tax code is complex. Congress could establish a carbon tax framework that rivals the complexity of a cap-and-trade program. For instance, a carbon tax that only applies to CO_2 emissions from fossil fuel combustion may be more transparent than a carbon tax that address non-CO_2 GHG emissions.

Moreover, policymakers could provide subsidies or exemptions to the fossil fuel industry that would run counter to a carbon tax. In addition, policymakers could allow for tax credits for carbon sequestration projects, similar to carbon offsets in a cap-and-trade regime. As with carbon offsets in a cap-and-trade program, this would require a further level of administrative responsibilities, and potentially weaken the program if the sequestration projects lack credibility.[56]

Ironically, transparency, particularly in regards to costs, could be a political liability for a carbon tax. Although both a carbon tax and a cap-and-trade program would impose higher energy costs, the costs from a cap-and-trade program would be more difficult to estimate, because the market would determine the price of emission allowances (and thus the overall costs of the program).

In other words, cap-and-trade proponents may have an advantage in promoting their approach, because they can cite studies that estimate relatively low program costs.[57]

Administrative Issues

A well-developed administrative structure for collecting taxes already exists in the United States. Moreover, fuel sales are well-documented and are currently taxed (for various reasons) to some degree.[58] Thus, a carbon tax on fossil fuels—if applied upstream in the economy (discussed below)—could utilize this existing framework. However, a carbon tax that included other GHG emissions or their inputs would not have the advantage of this existing administrative structure. Point of taxation issues are discussed later in this report.

Some argue that an emissions trading system would require the establishment of new institutions to administer the program.

> [T]he upfront costs of creating institutions to administer trading are significant and likely to produce entrenched bureaucracies that clamor for ever-tighter controls on carbon emissions.[59]

Others maintain that these concerns may be overstated. "Experience with existing trading programs, such as the U.S. SO_2 trading program, has shown that these institutions can arise quickly and for the most part inexpensively."[60] However, comparing the SO_2 trading program with a potential GHG emission trading program highlights appreciable differences. First, the scale of the SO_2 trading program is considerably smaller than the proposed cap-and-trade initiatives. Second, the SO_2 emission sources are more concentrated than GHG-emitting sources, which cut across multiple economic sectors.

As with other comparisons, the relative advantage of a carbon tax would depend on the designs of the instrument alternatives under scrutiny. For example, a cap-and-trade system with a more downstream structure may present more of a challenge, because more sources would be subject to the cap. In addition, if a cap-and-trade program were to provide emission allowances to covered sources at no cost, policymakers would need to devise a system for determining allotment. Emission allowance auctions would alleviate this responsibility, but would require an additional administrative responsibility, and some have raised concerns that auctions may impose a risk of market manipulation.

Another potential administrative advantage of a carbon tax approach is that it carries less risk of being hindered by litigation. Compared to a carbon tax, cap-and-trade legislation may delegate more authority to EPA or another government agency to implement certain components—e.g., specific allocation decisions, offset protocols, etc.—of the program.[61] Such decisions may be vulnerable to legal challenges from stakeholders.

Policy Modification

Some have argued that one of the advantages of a carbon tax is the relative ease—compared to a cap-and-trade program—in which the program's stringency could be modified.[62] In contrast, they assert that policymakers would face difficulties if they sought to adjust an emissions cap after the program's initiation.[63] The rationale for this assertion is that covered sources that made or purchased emission allowances beyond those needed in a given year would lose some of the value of these allowances if Congress raised (i.e., loosened) the cap at a later time. Similarly, a covered source may make capital investments based on the assumption of a stringent cap. If policymakers subsequently loosened the cap, these covered sources would take longer to recoup their investments.[64] However, this concern could also apply to a carbon tax. For example, energy producers and consumers may make investments based on an expected carbon tax. If the tax is subsequently altered, the value of such investments may change.

POTENTIAL DISADVANTAGES

A GHG emission reduction program that employs a carbon tax may face several disadvantages, compared to a cap-and-trade program. As with the potential advantages, the size and/or relevancy of the disadvantages are dependent on the design of the programs being compared. The potential disadvantages are discussed below.

Uncertain Emissions

The primary disadvantage of using a carbon tax to control GHG emissions is that the level of emissions would be uncertain.[65] For some this concern may present a non-starter, precluding a carbon tax as an option to control GHG emissions. As discussed above, some argue that the potential for irreversible climate change impacts necessitates the emissions certainty that is only available with a quantity-based instrument (e.g., cap-and-trade).

However, uncertain emissions do not necessarily equate with no emission control. Multiple models have estimated the carbon tax rate that would be required to achieve certain emission targets, targets that are comparable with the emission reduction goals of recent cap-and-trade proposals.[66] Regardless, these models can only provide estimates based on the best information available about the marginal costs of abatement (**Figure 1**). Although uncertain emissions are inherent with a carbon tax approach to emission control, there are policy options available to Congress to enhance the emission control certainty of a carbon tax. In theory, policymakers could devise a carbon tax program that yields only short-term emission fluctuations, as it progresses towards its long-term emission reduction objective. To achieve this goal, Congress would need to enact a mandatory GHG emission reporting regime to act in parallel with a carbon tax. Data from such a regime could be used to track the impact and performance of a carbon tax. If policymakers determine emission reduction is not occurring at a desired pace, the tax rate could be amended. As discussed above, carbon tax proponents would argue that short-term emission variations would not undermine efforts to control climate change. Indeed, they would assert that short-term emission fluctuations are greatly preferred to the price volatility that a cap-and-trade program would potentially impose. This argument is supported by the notion that CO_2 is a stock pollutant, which generates damages through its overall concentration in the atmosphere, not its annual flow. Critics of such an approach may question whether modifying the carbon tax rate on a periodic basis would be a politically palatable option. For example, it may be difficult for policymakers to increase the tax rate, especially during downturns in the economy. To address these concerns, some have suggested that Congress authorize an independent board or agency with the ability to modify the tax rate, in order to meet pre-determined emission reduction objectives.[67] Although this approach would improve emission certainty, long-term price certainty would be sacrificed to some degree, depending on the authority of the delegated entity to adjust the tax rate.

Political Feasibility

A common argument against a carbon tax is that it would present more political challenges to enact than an emissions cap, and thus policymakers should focus on the latter instrument. The Council on Foreign Relations recommended a cap-and-trade system over a carbon tax, stating:

> The Task Force finds that, assessed from a domestic economic perspective, the relative advantages of cap-and-trade and carbon tax approaches are easily overstated. Either, designed with a mix of efficiency and fairness in mind, would be an acceptable approach to reducing emissions.

> The Task Force finds, however, that the *political momentum* behind cap-and-trade makes its near term adoption much more likely.... [emphasis added][68]

Although a carbon tax would likely face more political obstacles than a cap-and-trade program, some of these obstacles may be based on misunderstandings regarding the differences between the two approaches or on assumptions that the tax would be set too low to be effective. Carbon tax proponents may be able to address these issues to some degree, but the recent political momentum for a cap-and-trade program may trump these efforts. This section examines some of the political concerns associated with a carbon tax.

What's in a Name?

Enacting a "carbon tax" to control GHG emissions would pose a political challenge for policymakers simply due to the word "tax." In contrast, the term "cap-and-trade" does not carry the same stigma. To some degree, the terminology advantage of a cap-and-trade program is arguably unfair, because it would act very much like a carbon tax. By placing a price on carbon, both a cap-and-trade and a carbon tax would likely impose costs on the economy. Indeed, some Members have argued that a cap-and-trade approach should be labeled as a "cap and tax," because it would act effectively as a tax.[69] Proponents of enacting a carbon tax understand the political ramifications of their instrument's name. To overcome this hurdle, some have described carbon tax programs as "user fees" or "tariffs" or if they are highlighting revenue distribution opportunities—carbon tax "swaps"[70] or "shifts." There may be procedural consequences of a name change—see the text box "The Role of Committee Jurisdiction," below. Proponents of a carbon tax could potentially make inroads in the debate by framing the debate within the larger context of tax policy, particularly tax reform.

Support from Industry?

Certainly the term "industry" covers a wide and diverse array of economic actors. Their motivations for support of, or opposition to, a carbon tax may vary as well. Support may depend on the anticipated design of the alternative—i.e., a cap-and-trade system. Cap-and-trade proposals have evolved in recent years. Industry stakeholders may have supported a cap-and-trade approach because, if enacted, they expected to receive some or perhaps all of the allowances at no cost.[71] This may not be an unrealistic expectation, considering the emission allocation strategies in the sulfur dioxide (SO_2) emissions cap-and-trade program and the European Union's Emissions Trading Scheme (EU ETS),[72] both of which provide almost all of their allowances to covered sources at no cost. The no-cost allowance allocation to EU emissions sources generated windfall profits for some industries.[73]

The Role of Committee Jurisdiction

The instrument's name and function would play an important role in terms of committee jurisdiction. A carbon tax proposal would likely be referred to the House Committee on Ways and Means and/or the Senate Committee on Finance; a cap-and-trade proposal would likely be referred to a committee that covers environmental policy (e.g., the Senate Committee on Environment and Public Works, House Committee on Energy and Commerce). In general, environmental policy committees may have different perspectives, expertise, or priorities than tax committees. These differences could influence the design of a GHG control instrument that originates in a particular committee, as well as subsequent program oversight and adjustments: e.g., the use of revenue from a carbon tax or a cap-and-trade system. The committee jurisdiction factor may play some role in the conceptualization of, and the debate over GHG emission control.

Some may contend that if the policy were called a "user fee" or "user charge" then the primary committee of jurisdiction would not necessarily be a tax policy committee. Standing committees of the House, other than the Appropriations and Budget Committees, may report legislation creating or modifying user fees. However, this procedural point would likely be the subject of intense debate: A ruling on whether a policy instrument is a user fee or a tax measure may depend on the nature of the charge rather than its label. For more information on this issue see U.S. Office of Management and Budget, Analytical Perspectives, Budget of the United States Government, Fiscal Year 2009 (2008); and Congressional Budget Office, The Growth of Federal User Charges (1995).

Some Members have argued that a cap-and-trade proposal would ultimately act as a tax on GHG emissions. They question whether the potential of a cap-and-trade program to raise revenue would have implications for committee jurisdiction. See Letter from Representatives Sensenbrenner and Cantor to Representatives Conyers and Nadler (February 11, 2009), at http://republicans.globalwarming.house.gov/index.shtml.

However, the more recent cap-and-trade proposals from the 110th Congress would have provided considerably fewer allowances to covered sources at no cost.[74] Industry stakeholders have voiced concern over the apparent evolution of allocation strategy.[75] If industry expects to receive fewer allowances at no cost than previously thought, their support of a cap-and-trade system may erode, which could increase interest in a carbon tax.

Some argue that if industry is acting in its own self-interest, taxes would be the least-preferred option, even behind auctioned allowances in a cap-and-trade system.[76] The explanation for this preference is that auctions may allow room for industry to manipulate the price, driving down the overall cost of compliance (as opposed to a tax, which would be fixed). But some industries would prefer the cost to be fixed and known in advance (as a carbon tax would do), so they can better assess investment options (e.g., equipment upgrades, efficiency improvements, etc.).[77] In addition, some industry stakeholders may lend their support to a tax, if the alternative were to resemble recent cap-and-trade proposals (from the 110th Congress), which would have applied directly to upstream entities (i.e., energy producers). This approach would resemble a tax—without a price certainty—for downstream energy consumers (e.g., power plants).[78] Moreover, the upstream entities subject to direct compliance—i.e., required to submit allowances—would have little or no direct emissions. The gap between compliance and emissions could pose challenges in a situation without a known price (i.e., a cap-and-trade program).

Support from Environmental Groups?

Supporters of stringent GHG emission control may oppose a carbon tax, because they assume that if Congress enacts a carbon tax, Members would set the rate too low to induce emission mitigation: e.g., energy conservation, fuel substitution, or technology development. This may be more of an argument against the tax rate, than the instrument itself.

A similar concern of stringency may exist if policymakers decide to employ a cap-and-trade system. For example, the cap might be set at undesired levels, or the program may allow for the use of questionable offsets, which could pose a risk of undermining the objective of emission reduction. Likewise, the pressure to delay the strengthening of an established cap that is considered ineffective or expensive could be considerable. Regardless, an emissions cap—at least on paper—offers more of a guarantee that a specific emission level would be met. Some environmental groups may assert that this perceived guarantee overshadows other concerns.

Consideration of International Efforts and Cooperation

It is generally recognized that mitigation of human-induced climate change will ultimately require action by all major emitting countries. Therefore, when assessing different mechanisms of GHG emission reduction, Congress may want to consider how an market-based instrument—e.g., a cap or a tax—would 1) complement existing or future international efforts and/or 2) maximize participation among major emitting nations.

For policymakers and negotiators who are attempting to create an international GHG reduction regime, establishing either instrument on an international scale would present substantial and unique challenges. A carbon tax may be at a disadvantage, primarily because the most prominent international activities currently involve cap-and-trade programs. However, some assert that a carbon tax approach may present a stronger opportunity to maximize participation between the United States and developing economies—e.g., China and India—that emit substantial portions of GHG emissions. Some of the issues are discussed below.

Coordination with Existing International Efforts

At the core of the international GHG mitigation efforts underway—e.g., the Kyoto Protocol and the EU ETS—is a cap-and-trade program among nations with quantitative commitments. If Congress enacted a cap-and-trade system in the United States, the program could be designed to allow trading between systems. In terms of overall economic efficiency, linked trading programs would be ideal, because covered sources would have access to more emission reduction opportunities. However, linked programs could yield unintended consequences if they differ in scope and/or stringency.[79]

A U.S. carbon tax program could conceivably have a linkage of sorts with international cap-and-trade programs. For example, a carbon tax system could allow for tax credits for emission reductions made outside of the United States.[80] However, this trade would only flow in one direction: Covered sources in the EU could not purchase reductions from covered sources in the United States. Thus, this type of linkage would not equate with a union of cap-and-trade programs. Sources in the United States that were subject to the carbon tax would find this opportunity worthwhile only if the carbon tax rate were higher than the emission allowance price (EU emission allowances or EUAs) or, if eligible, offset project prices (traded as certified emission reductions or CERs) in the EU ETS. However, if policymakers were to provide tax credits for international emission abatement projects, concerns similar to those involving offsets in a cap-and-trade system would be raised. If the illegitimate projects qualify as tax credits, the primary objective of the program—GHG emission control—may be compromised.

Maximizing Participation

Some have raised the argument that encouraging maximum participation among nations should be a higher priority than instrument choice and the potential consequences of choosing a cap over a tax (e.g., loss of economic efficiency).[81] Would a carbon tax or a cap-and-trade program be the most effective approach to bring the most nations, or more precisely, emissions, under a mitigation umbrella? Neither strategy would necessarily require equal commitments between nations. For example, the United States and China could each establish unilateral carbon tax programs—as opposed to an international, harmonized system—with different tax rates. Indeed, an international carbon tax system would encounter substantial implementation challenges, particularly regarding issues of national sovereignty.

It would be (and has proven) difficult to achieve wide and diverse participation—namely, agreement from both the United States and developing nations—in a GHG control regime. Some may favor a cap-and-trade approach, because of the buildup of existing programs and the political momentum behind the cap-and-trade approach. However, this reasoning may be questioned on several counts. First, although a substantial percentage of the world's CO_2 emissions are covered under the Kyoto or EU ETS cap, an even larger percentage are not covered.[82] This latter group includes the United States, China, and India, which accounted for approximately 54% of the CO_2 emissions from the top-25 CO_2 emitting nations in 2005.[83] Second, the mere existence of a program (and its political momentum) does not indicate it is the optimal solution. Indeed, the existing international cap-and-trade systems, which are still in their early stages, have encountered substantial problems.[84]

However, a carbon tax approach would encounter obstacles similar to those observed in the cap-and-trade discussions. For example, it would be difficult to reach agreement on the equitable or fair tax rates for each nation. Not only would this agreement be difficult between developed and developing nations—for multiple reasons that are beyond the scope of this report—but it would also pose a challenge between developed nations. For example, a 2000 study found that European nations were (at the time of the study) effectively taxing carbon at a rate of approximately $27/ton of CO_2.[85] If European nations and the United States sought to establish their own carbon tax rates, pre-existing carbon taxes (and subsidies) may complicate the discussions.

International Implementation Concerns

Regardless of the chosen instrument, either would impose its own set of logistical challenges. Some assert that an international cap-and-trade system would be more susceptible to corruption.[86] However, it could be argued that monitoring and enforcement of carbon taxes in individual nations might pose similar difficulties. A primary obstacle would be determining a nation's "net carbon tax," because nations or political subdivisions could offset the carbon tax through energy subsidies or loopholes for the entities most impacted by the tax.[87] Moreover, some nations may not have transparent tax systems in place now. Adding a carbon tax to such a framework could be problematic.

Implementation of a Carbon Tax

If Congress were to use a carbon tax to control GHG emissions, policymakers would face several implementation decisions: (1) the point of taxation—where to impose the tax and what to tax; (2) the level of taxation; and (3) how to apply the tax revenue. These considerations are discussed below.

Point of Taxation

A point of taxation debate is analogous to a debate over the scope or coverage of a potential cap-and- trade program. Congress would face similar considerations of emissions coverage with either instrument. Describing coverage issues, EPA stated:

> Ideally, all sources, sectors, and emissions would be included for full coverage and maximum environmental effectiveness and economic efficiency. However, measurement capabilities and costs, available control options, administrative burdens, political considerations, and other constraints may limit participation to a subset of emission sources.[88]

A comprehensive discussion of these issues is beyond the scope of this report,[89] but the following sections provide options for policymakers. Note that these sections overlap to some degree.

Where to Impose a Carbon Tax?

Regarding where in the economy to impose a carbon tax, the options—at least for fossil fuel combustion-related CO_2 emissions[90]—are often couched in terms of their location in the energy production and consumption chain: upstream or downstream (**Figure 7**):

- **Upstream**—An upstream approach would apply a carbon tax to fossil fuels when they enter the U.S. economy, either at the mine, wellhead, or another practical "chokepoint" in the production chain. Imported fuels would be addressed at their point of entry into the United States (**Figure 7**). Under this approach, a carbon tax would be levied before emissions have occurred. The tax would be based upon the carbon content of the fossil fuel produced. An advantage to the upstream approach is that the number of sources directly subject to the tax would be minimized, while covering a broad spectrum of the economy.
- **Downstream—**A downstream approach would apply the tax on the emissions themselves. For example, emission sources, such as power plants or industrial facilities,[91] would pay a fee for each ton of CO_2 released into the atmosphere. In some economic sectors, there may be advantages to imposing a carbon tax downstream, even if more entities would be directly subject to the tax. For instance, Congress may consider a downstream approach for the electric utility industry, because power plants already continuously monitor their CO_2 emissions to the EPA.[92] There may be other advantages to applying the tax directly to emissions from coal-fired power plants. For example, some argue that this approach would provide a greater stimulus[93] to the development of carbon capture and sequestration (CCS) technology.[94] However, in some sectors—e.g., transportation or residential—a downstream approach may be challenging, and perhaps impractical to implement because of the comparably large number of emission sources.
- **Combined Approach**—Sometimes referred to as a hybrid strategy, this approach would cover emissions from different economic sectors in different manners. For example, transportation fuels could be taxed upstream at the point of extraction (or midstream at the refinery); coal could be taxed at the emission sources: power plants and specific industrial sectors.

Where Congress decided to impose a carbon tax would determine which entities would be required to (1) make tax payments based on emissions or emission inputs, (2) monitor emissions or emission inputs, and (3) maintain records of relevant activities and transactions.

A direct tax on emissions may be easier to understand, but considerably more difficult to implement across all economic sectors, due to the large number of emission sources. A complicating factor is that some fossil fuel uses—e.g., feedstock in asphalt production—do not generate emissions but sequester the carbon. Policymakers may consider mechanisms to exempt the fuels used for such purposes or allowing for tax credits for these activities.

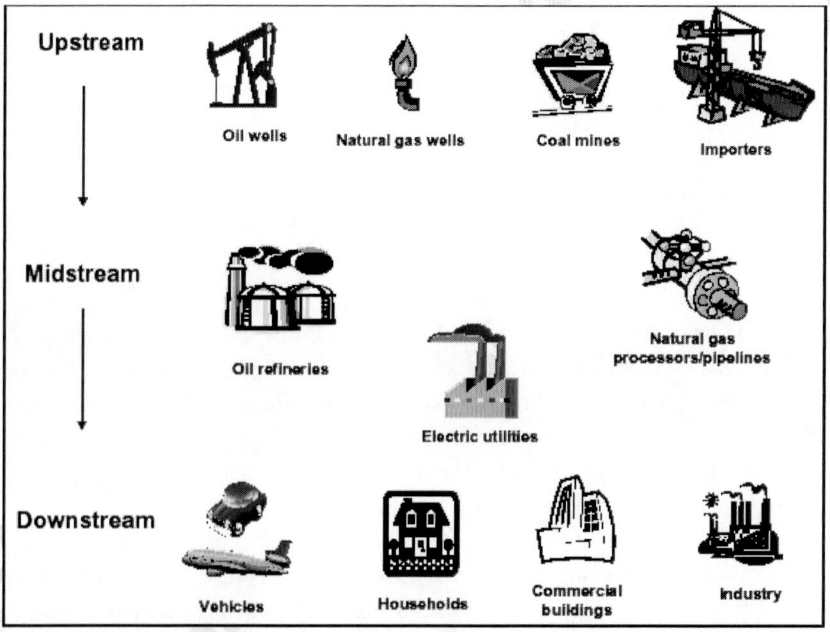

Source: Prepared by CRS.

Note: Electric utilities could be listed as either downstream entities—because they are direct sources of emissions—or midstream, because their emissions are tied to the electricity consumption of their customers, the further downstream consumers.

Figure 7. Illustration of Options for Points of Taxation within the Energy Production-to-Consumption Chain.

Table 1. CO_2 Emissions Per Unit of Energy for Fossil Fuels

Fossil Fuel	CO_2 Emissions Per Unit of Energy (million metric tons/quadrillion BTU)
Coal	94
Crude oil	74
Natural gas	53

Source: Prepared by CRS, based on CO2 emission factors from EIA, at http://www.eia.doe.gov/environment.html.

Note: CO_2 emissions from coal vary by the type of coal being used.

Fossil fuels have varying levels of carbon content, thus a carbon tax would impact fossil fuels differently. For example, coal has almost twice the carbon content per unit of energy as that of natural gas (**Table 1**). Thus, a carbon tax would raise the price of coal more than the price of natural gas.

CO_2 Emissions or All CHG Emission?

Another point-of-taxation issue is whether a carbon tax should cover CO_2 emissions or a more comprehensive array of GHG emissions. Mitigation opportunities from non-CO_2 gases are often less expensive than other reduction options. Economic studies indicate that one advantage of including more or all GHG emissions, as opposed to only CO_2 emissions, is that the overall costs of the GHG reductions would be decreased.[95] A recent analysis of a carbon tax proposal—H.R. 3416 (Larson)—from the 110th Congress found that by adding other GHG emissions to the tax base, the overall costs of achieving the same amount of emission reduction would fall by 20%.[96] However, this finding was based on the inclusion of GHG emissions from all sources. As discussed below, the majority of the non-CO_2 GHG emissions come from sources that may present tax implementation challenges.

The 1993 Energy or BTU Tax

A carbon tax is different from an energy consumption tax, like the British thermal unit (Btu) tax that was proposed by President Clinton in 1993. Compared to a carbon tax, a Btu tax would be a less efficient mechanism to reduce CO_2 emissions. As highlighted in, fossil fuels generate different amounts of CO_2 emissions per unit of energy. A Btu tax would treat energy sources equally, thus a Btu tax would serve different objectives than a carbon tax.

> The 1993 Btu tax proposal called for a levy of 25.7 cents per million Btu, with a surcharge of 34 cents/million BTU on petroleum. The goals of the 1993 Btu tax proposal were to promote energy conservation and raise revenue. At the time, the proposed tax would have generated a new revenue stream of about $30 billion per year. The proposal was met with strong opposition and was not enacted; Congress ultimately enacted an (approximately 5-cent) increase in the motor fuels taxes.

Which Emission Sources to Control?

GHG emissions are spread throughout the economy and are generated by millions of discrete sources: e.g., smokestacks, exhaust pipes, households, and livestock. Although CO_2 is the primary GHG, some GHG emissions sources predominantly emit non-CO_2 GHGs. When determining which sources and gases to control through a tax, policymakers would need to balance the benefits of inclusion with administrative costs.[97] To achieve a specific emissions target, a tax applied on only a subset of economic sectors—thus covering only a percentage of U.S. GHG emissions—would need to have a higher rate than a tax applied to all GHG emissions. Numerous rationales have been put forward to justify excluding one group or sector from a reduction requirement, or to provide some other special consideration.

Table 2 lists the top emission sources of six GHGs in the United States. These sources combine to account for approximately 95% of the total (based on 2006 data).[98] **Table 2** also provides potential points of taxation that would cover the emissions from these sources.

CO_2 emissions from fossil fuel combustion yielded approximately 80% of U.S. GHG emissions in 2006.[99] Policymakers could address CO_2 emissions from fossil fuel combustion and non-energy uses—in aggregate 82% of U.S. GHG emissions—by levying an upstream carbon tax on fewer than 2,500 upstream or midstream entities (**Table 2**). A portion of these entities are already subject to environmental taxes that support other objectives.[100]

A downstream approach may be feasible for electric generating facilities powered by fossil fuels, which number approximately 3,300,[101] but impractical for other fossil fuel combustion sources— automobiles, offices, homes—which number in the millions. One possible alternative to a complete upstream/midstream strategy would be to address coal, the most carbon-intensive fossil fuel, at its combustion location. In 2006, coal-fired power plants accounted for 92% of U.S. coal combustion.[102] As shown in **Table 2**, there are about 641 coal-fired power plants in the United States. Some argue that there would be advantages to applying a point of taxation at electric generating facilities. For example, these facilities may have a stronger incentive and/or capability to seek tax credit opportunities, such as biological sequestration projects (if allowed within the tax system). Moreover, some contend that the development of CCS would be enhanced if coal-fired facilities were directly subject to the tax. The rationale for this argument is that under an upstream approach (that indirectly taxes coal-fired emissions) the price signal may be weaker than if the tax were applied directly to emissions at coal-fired facilities. However, other sources of coal combustion would need to be covered to avoid leakage situations—i.e., increased coal (and thus emissions) use in the uncovered sectors. It may be burdensome to address the remaining sources of coal combustion at the point of emissions, particularly at residences and commercial entities.[103] Beyond fossil fuel combustion, policymakers could cover more GHG emissions by expanding the coverage to some or all of the remaining sources listed in **Table 2**. Individually, many of these sources account for relatively small proportions of U.S. GHG emissions, but if some combination of these sources were included in the carbon tax base, the increase in emissions coverage could be appreciable. The remaining source categories vary by (1) the number of entities that would be subject to a carbon tax, and (2) the circumstances of emission release. In general, a large number of entities subject to a carbon tax would present higher administrative costs per unit of emissions. Regardless of entity number, some emissions sources would be more difficult to cover because of logistical challenges, such as emission measurement and/or monitoring. Although GHG emissions from the agriculture sector—farming and livestock—account for approximately 6% of total GHG emissions in the United States, agricultural emissions would be comparatively difficult to cover under a carbon tax. As indicated in **Table 2**, these emissions are generated by a large number of entities, and there are no upstream measurement points analogous to those that could cover emissions from the transportation sector. However, nitrous oxide (N_2O) emissions from the use of artificial fertilizers—as opposed to natural fertilizers like manure—might be an emission source that could be addressed through an upstream point of taxation. EPA estimates that approximately 25% of the N_2O emissions from agricultural soils are related to artificial fertilizer use.[104] This represented approximately 0.8% of U.S. GHG emissions in 2006, so it would be included in **Table 2** if counted separately. A tax on fertilizers may lead to less fertilizer use, but some argue that it could stimulate other activities—e.g., increased use of natural fertilizers—that would release

N_2O emissions (a form of emissions leakage).[105] This issue is beyond the scope of this report.

Table 2. Selected Sources of U.S. GHG Emissions and Potential Applications of a Carbon Tax

GHG Emission Source	Percentage of U.S. GHG Emissions (2006 data)	Potential Carbon Tax Applications	
		Entity	**Number**
CO_2 from fossil fuel combustion:	79.9	Coal mines[a]	1,374
-electricity generation		or	641
- transportation		Coal-fired power plants[b]	
- industrial		Power plants[c] using imported coal[d]	26
- commercial/residential			
		Petroleum refineries[e]	150
		Petroleum importers[f]	220
		Natural gas processors[g]	530
		Natural gas importers[h]	45
N_2O from agricultural soils	3.8	Farms[i]	>2million
CO_2 from non-energy use of fuels	2.0	Covered by the tax applied to fuels above)[j]	
CH_4 from livestock (enteric fermentation)	1.8	Cattle operations[k]	967,440
CH_4 from landfills	1.8	Landfills[l]	1,800
HFCs from the substitution of ozone depleting substances	1.6	HFC manufacturers[m]	5
CH_4 from natural gas systems	1.5	Natural gas processors	530
CH_4 from coal mines	0.8	Coal mines[n]	1,374
CO_2 from iron/steel production	0.7	Raw steel production facilities;	116
		Integrated steel mills[o]	18
CO_2 from cement manufacturing	0.6	Cement plants[p]	118
CH_4 from manure management	0.6	Cattle operations;	967,440
		Swine operations[q]	65,640
Percentage of Total GHG Emissions	95.1		

Source: Prepared by CRS; GHG emission data from EPA, EPA, *Inventory of U.S. Greenhouse Gas Emissions and Sinks: 1990-2006* (April 2008); data for number of entities from multiple sources, cited in footnotes.

a. This figure accounts for mines in operation in 2007. EIA, *Coal Production and Number of Mines by State and Mine Type*, (2008).

b. In 2006, there were 641 plants with at least 1 coal-fired generator unit (EIA, 860 Database).

c. Number of plants comes from EIA database, Monthly Nonutility Fuel Receipts and Fuel Quality Data (Database 423). CRS was unable to determine the number of companies that act as coal importers, analogous to petroleum importers.

d. In 2006, the United States imported approximately 36 millions short tons of coal (EIA, *Quarterly Coal Report* (2008), table 4)—3.5% of the amount to coal consumed domestically in that year (EIA, *Annual Coal Report 2006* (2007), table 26). Coal imports have increased by more than 200% since 2002.

e. This figure represents the number of "operable" refineries. EIA, *Refinery Capacity Report* (2008).

f. EIA, Company Level Imports (as of November 2008). Note that some of these companies may import only crude oil, whose emissions would be covered by the tax at the domestic refineries. Thus, this figure represents an upper bound of petroleum product importers potentially subject to a carbon tax.

g. EIA, Natural Gas Processing: The Crucial Link Between Natural Gas Production and Its Transportation to Market (2006).

h. This includes pipelines and liquefied natural gas facilities. EIA, *About U.S. Natural Gas Pipelines* (as of September 2008).

i. U.S. Department of Agriculture, *Farms, Land in Farms, and Livestock Operations: 2007 Summary* (2008). The resource defines a farm as "any place from which $1,000 or more of agricultural products were produced and sold, or normally would have been sold, during the year."

J. Fossil fuels are used for a wide range of non-energy purposes. EPA estimates that of the total carbon consumed for non-energy purposes, approximately 62% is stored in products, and not released to the atmosphere (EPA, *Inventory of U.S. Greenhouse Gas Emissions and Sinks: 1990-2006* (April 2008), tables 3-14 and 3-15). The 2% value in Table 4 represents the emissions. In an upstream carbon tax system, fuels would be taxed before they are used. Congress may want to consider providing tax credits for the amount of carbon stored in products.

k. U.S. Department of Agriculture, Farms, Land in Farms, and Livestock Operations: 2007 Summary (2008).

l. EPA, Inventory of U.S. Greenhouse Gas Emissions and Sinks: 1990-2006 (April 2008), citing BioCycle, 15th Annual BioCycle Nationwide Survey: The State of Garbage in America (2006).

m. Intergovernmental Panel on Climate Change, Safeguarding the Ozone Layer and the Global Climate System, Issues related to Hydrofluorocarbons and Perfluorocarbons (2005), Figure 11.1.

n. Methane from underground mines, which accounts for about 61% of coal mine methane, is removed through ventilation systems for safety reasons. These emissions would be easier to monitor under a carbon tax than aboveground coal mine methane emissions.

o. Data from U.S. Geological Survey, Mineral Commodity Summary, Iron and Steel Production (2008), at http://minerals.usgs.gov/minerals/pubs/commodity/iron_&_steel/.

p. Cement manufacturing information from Portland Cement Association, at http://www.cement.org/basics/cementindustry.asp.

q. U.S. Department of Agriculture, Farms, *Land in Farms, and Livestock Operations: 2007 Summary* (2008). Other animals—chickens, horses, and sheep—contribute approximately 10% of the total emissions from manure (EPA, *Inventory of U.S. Greenhouse Gas Emissions and Sinks: 1990-2006* (April 2008), table 6-6).

In many cases, the emissions from agricultural activities may be difficult to monitor and/or measure. For practical concerns and other reasons—e.g., political considerations—the agriculture sector has thus far been excluded as a capped source in the cap-and-trade proposals that have specifically designated covered sources.[106] If not subject to a carbon tax, agriculture emissions could be addressed with other policy tools. For example, tax credits could be provided for qualifying reduction, abatement, or sequestration projects.[107]

The other emission sources listed in **Table 2** may be more amenable to a carbon tax, at least in terms of the number of entities and emission circumstances. Compared with agricultural sources, the other emission source categories—excluding HFCs (discussed below)—comprise a relatively small number of entities, at which emissions are easier to measure and monitor. In fact, certain sources—e.g., some landfills and underground coal mines—currently control their (methane) emissions for air pollution and safety reasons.

Methane from landfills—1.8% of U.S. emissions—could be covered by including an additional 1,800 landfills in the tax base. To some degree, large municipal landfills are already required per federal air regulations to collect and combust methane emissions. Other landfills are following suit under state laws or voluntary programs.[108]

Hydrofluorocarbons (HFCs) are used as substitutes for several classes of ozone depleting substances (ODSs—hereafter, HFC-ODS).[109] HFC-ODS account for 1.6% of total U.S. GHG emissions (**Table 2**). HFC-ODS emissions occur under different circumstances than the other, "smaller" sources discussed above—landfills, coal mines, etc. Most of the HFC emissions (86% in 2006) come from air conditioning and refrigeration units—located in vehicles, buildings, and homes—during production, operation, and disposal.[110] Unlike the other "small" source categories, which emit GHGs onsite, HFC-ODS emissions enter the atmosphere from millions of locations, far away from and long after HFC production. Because of the different emission circumstances of HFC-ODS, some have proposed combining a tax at HFC production—five facilities in the United States—with a tax credit for HFCs properly recycled or disposed.[111] For example, automobile air conditioners could be sent to designated sites in order to recycle or dispose of the HFCs. Norway has an HFC rebate program in place that may be instructive for U.S. policymakers.[112] As with other carbon-intensive materials, Congress may consider applying a tax to imported HFCs as well.

By applying the tax to iron/steel production and cement manufacturing facilities—approximately 250 entities—an additional 1.3% of U.S. emissions could be covered.[113] A further 1.5% could be addressed by including methane from natural gas systems. Roughly 25% of this amount comes from field production, which may be impractical to monitor and measure accurately.[114] The remainder involves emissions from normal operations, routine maintenance, or "system upsets." Fugitive emissions (or accidental releases) are the primary source of emissions from these activities.[115] A tax on these emissions may impose administrative costs in terms setting up a process to measure and monitor these emissions.

Methane emissions from underground coal mines—0.8% of total GHG emissions—may be easier to monitor under a tax than at aboveground mines. At all underground mines, methane is ventilated for safety reasons; methane from aboveground operations is released directly to the atmosphere. The 0.8% figure includes methane only from underground coal mines, which account for about 61% of methane from coal mines.[116]

Level of Taxation

If Congress were to use a carbon tax to control GHG emissions, setting the level of the carbon tax would be a subject of intense debate. Regardless of the chosen tax schedule, some parties would argue that the selected rate was either too low or too high. Although the decision would likely involve political calculations, there are economic principles that could be used to inform the debate. This section identifies and discusses two approaches that policymakers could take to set the level of a carbon tax. The first approach is based on the estimated benefits associated with avoiding climate change impacts. The second approach is based on a GHG emissions target and the tax rate needed to achieve that target. The first approach is arguably stronger from a theoretical standpoint, but the second approach is arguably (considering the uncertainty associated with the first approach) a more practical option.

To provide some context for potential carbon tax rates, **Table A-1** of this report (located in the **Appendix**) includes estimates of price increases to fossil fuels and motor gasoline based on different carbon tax rates.

Tax Based on Estimates of Costs and Benefits[117]

As noted previously, economic principles involving pollution taxes prescribe that the optimal tax should be at the intersection of the marginal costs of emission abatement with the marginal benefits of abatement. In terms of climate change, the intersection of costs and benefits is often described as the Social Cost of Carbon (SCC). The Intergovernmental Panel on Climate Change (IPCC) defines the SCC as the following:

> The discounted monetized sum (e.g. expressed as a price of carbon in \$/tCO_2) of the annual net losses from impacts triggered by an additional ton of carbon emitted today. According to usage in economic theory, the social cost of carbon establishes an economically optimal price of carbon at which the associated marginal costs of mitigation would equal the marginal benefits of mitigation.[118]

For multiple reasons, applying this economic principle to set a carbon tax currently presents monumental, if not insurmountable, challenges. These are highlighted below.

Social Cost of Carbon Estimates

Estimates of the SCC are fraught with uncertainty, thus the wide range of estimates produced should not come as a surprise. According to a 2008 meta-analysis, there are over 200 SCC estimates from 47 studies. The estimates range across several orders of magnitude, from zero to over \$500 per ton of CO_2-equivalent (tCO_2-e) emissions.[119] The uncertainty and controversy of choosing a value within this range would be formidable.

The definition of SCC indicates the complexity involved in its estimation. The SCC can be defined as a monetary measure of the marginal damages (or costs)—on a global scale—of an incremental unit (e.g., one metric ton) of anthropogenic GHGs emitted to the earth's atmosphere at some point in time (e.g., now).

Estimating the SCC requires analysts to place values on goods and services that may be difficult (or controversial) to precisely measure: e.g., human health/life, water supplies, agricultural production, recreational activities. The value of these goods/services will vary from one location to the next, complicating the estimate. Moreover, the climate change processes that may impact these goods and services are not completely understood.[120]

The element of time particularly complicates an SCC estimation and application in several ways. An SCC estimate signals—per economic theory—what global society should be willing to pay *now* to avoid *future* damages due to additional emissions generated *today*. As discussed above, CO_2 and other GHGs remain in the atmosphere for decades to centuries, thus current emissions could generate damages for years to come. To take time into account—i.e., determine the cost of future damages in today's dollars (referred to as net present value)—economists employ a discount rate.[121] Discount rates are typically at the crux of cost-benefit analyses and their assignment often generates controversy. As with other public policy decisions that employ costbenefit analysis, the chosen discount rate is a critical factor when estimating the SCC. In terms of uncertainty in evaluating damages from climate change, the Intergovernmental Panel on Climate Change (IPCC) estimates that the discount rate is the second most important factor (climate sensitivity assumptions are number one).[122]

An additional factor included in SCC estimates is the assumption of the current global emission trajectory. As stated in the Stern Review, "an SCC curve cannot be drawn, nor an SCC calculated, without specific assumptions on future paths."[123] A business-as-usual (BAU) path assumption will produce a higher SCC estimate than an assumed path of emission stabilization.[124] A higher concentration pathway yields a higher SCC estimate, because if an extra ton of GHG emissions is added to an atmosphere with a high GHG concentration, the extra ton is projected to yield more damages than if it were added to an atmosphere of lower GHG concentration.[125]

Applying Social Cost of Carbon Estimates

Using SCC estimates as a basis for a carbon tax would be consistent with economic principles, but extremely difficult in practice. The primary reason is that the estimates vary widely, based on the underlying assumptions used to generate the estimates. Selecting one estimate over another may be difficult to justify. For these reasons, economic reports often discuss the SCC in terms of a "starting place"[126] or a "good guide"[127] for setting a carbon tax level.

Moreover, the estimates are snapshots for a given point in time. As conditions change, the SCC estimates change. Thus, using them as a basis for a carbon tax rate may be problematic, because policymakers would need to alter the tax schedule based on the most recent estimates. However, this concern would exist for a carbon tax system based on any economic estimates, such as marginal cost curves (as discussed below).

Regardless, the tax rates from several carbon tax proposals in economic literature[128] and one proposal from the 110th Congress—H.R. 3416 (Larson)—are close in value to the mean (average) of SCC estimates identified in a recent meta-analysis.[129] In 2007, the IPPC found that *peer-reviewed* estimates[130] of the SCC in 2005 dollars have an average of \$12/tCO_2-e.[131] However, prominent economists have generated estimates above and below this level.[132]

Tax Based on Meeting an Emissions Target

An alternative to the above approach—basing a carbon tax on expected damages (measured in net present value)—would be to determine the tax rate needed to stimulate a specific emissions reduction goal. For example, Congress would set an emissions target—e.g., 66% below 2005 levels by 2050[133]—and estimate the tax schedule—e.g., X/tCO_2$-e in 2010, increasing Y% per year—that would encourage the reductions needed to meet the target. Using this approach, policymakers would not need an estimate of the marginal benefits of abatement in the tax rate calculation. However, the chosen emission reduction target may involve some debate over the level of climate change risks that could be tolerated by society.[134]

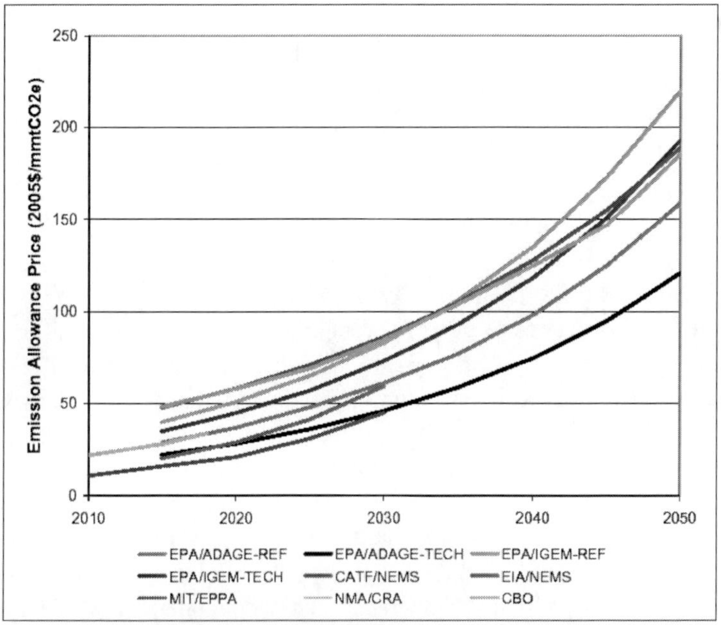

Sources: (1) EPA/ADAGE and EPA/IGEM: "Data Annex" available on the EPA website at http://www.epa.gov/climatechange/economics/economicanalyses.html;

(2) MIT/EPPA: Sergey Paltsev, et al., "Appendix D" of Paltsev et al., Assessment of U.S. Cap-and-Trade Proposals, MIT Joint Program on the Science and Policy of Global Change (2007);

(3) EIA/NEMS: EIA, Energy Market and Economic Impacts of S. 2191, the Lieberman-Warner Climate Security Act of 2007 (April 2008);

(4) CATF/NEMS: Jonathan Banks, Clean Air Task Force, The Lieberman-Warner Climate Security Act—S. 2191: A Summary of Modeling Results from the National Energy Modeling System (February 2008);

(5) NMA/CRA: CRA International, Economic Analysis of the Lieberman-Warner Climate Security Act of 2007 Using CRA's MRN-NEEM Model (April 8, 2008);
(6) CBO: Congressional Budget Office, Cost Estimate: S. 2191: America's Climate Security Act of 2007 (April 10, 2008).

Notes: This figure was adapted from CRS Report RL34489, *Climate Change: Costs and Benefits of S. 2191/S. 3036*, by Larry Parker and Brent D. Yacobucci. Estimates extrapolated by CRS from available data where necessary. Estimates converted to 2005$ using GDP implicit price deflator. The original figure included two additional estimates that are not included above, because these estimates were subsequently found to contain errors in the underlying methodology.

Figure 8. Emission Allowance Price Estimates under S. 2191

This approach might require estimates of the marginal costs of emission abatement. These estimates, while not as uncertain according to many economists, are similarly based on multiple assumptions, and are thus also fraught with substantial uncertainty. A comparison of the multiple cost analyses of the Lieberman-Warner cap-and-trade proposal (S. 2191/S. 3036) from the 110th Congress demonstrates the uncertainty in estimating costs.[135] These analyses forecast emission allowance prices within the narrow parameters of S. 2191. Although not the same measure, emission allowance prices provide an approximate value of marginal abatement costs for that proposed strategy. As shown in **Figure 8**, emission allowance price estimates for S. 2191 varied considerably, from \$16/tCO2-e to \$49/tCO_2-e in 2015. The figure indicates that the range of estimates broadens as time progresses.

The range of emission allowance prices—a proxy for marginal abatement costs—indicates that policymakers would have weak information on which to set the optimal tax rate to meet a selected emissions target. However, an uncertain emissions pathway goes hand-in-hand with a carbon tax system. As discussed above, this problem could be alleviated by annually monitoring U.S. emissions and altering the tax rate as deemed necessary—thereby requiring Congress or a delegated entity to adjust the tax rate through subsequent legislation or regulation.

Tax Revenue Distribution

Carbon tax revenues would be substantial, and policymakers could apply them to further multiple objectives (not listed in any particular order): allow for reductions in other taxes; alleviate the burden of increased energy prices (brought on by the carbon tax); stimulate development of low-carbon technologies; and support other climate- or non-climate-related purposes. The redistribution of carbon tax revenues is often described as "revenue recycling."[136] After providing an estimate of potential tax revenues for context, this section discusses some of the revenue recycling options and their potential impacts.

Estimates of Tax Revenues

Table 3 provides estimates of tax revenue estimates from recent legislative proposals from the 110th Congress: H.R. 2069 (Stark) and H.R. 3416 (Larson). Both bills would have imposed a carbon tax on coal, petroleum, and natural gas, based on the CO_2 emissions[137] the fossil fuels would generate upon combustion. As indicated in **Table 3**, the bills were estimated to generate between \$69 billion and \$126 billion in 2015. To put these estimated tax revenues in context, consider the federal net tax revenue from the three largest revenue sources for FY2005 (in 2005 dollars):[138]

- individual income tax: \$1,107 billion;
- employment taxes: \$771 billion;
- corporate income taxes: \$307 billion.

Use of Tax Revenues

Revenues from the carbon tax could be used to promote a wide range of policy objectives, climate-related or otherwise. This section identifies different applications of the carbon tax revenues. A subsequent section discusses the policy considerations for Congress surrounding the different applications of carbon tax revenue.

Table 3. Estimates of Potential Tax Revenues from Carbon Tax Proposals from the 110th Congress (in 2005 dollars)

	Proposed Tax Rates		Annual Tax Revenue Estimates	
Carbon Tax Proposals	**in 2015**	**in 2030**	**in 2015**	**in 2030**
H.R. 2069 (Stark)	\$10/tCO2-e	\$43/tCO2-e	\$69 billion	\$263 billion
H.R. 3416 (Larson)	\$20/tCO2-e	\$83/tCO2-e	\$126 billion	\$361 billion

Source: Prepared by CRS with estimates from Gilbert Metcalf *et al.*, *Analysis of U.S. Greenhouse Gas Tax Proposals* (2008), MIT Joint Program on the Science and Policy of Climate Change, Report No. 160.

Note: The MIT model assumed that H.R. 2069 and H.R. 3416 would start in 2012, although both bills (as written) would have begun in 2008.

Offset Reductions in Other Taxes

Carbon tax revenues could be used to replace taxes that apply to desirable activities (as noted earlier). Economic theory generally supports a tax policy that would reduce taxes on favored activities (e.g., increased employment or investment) and increase taxes on less desirable behavior (e.g., increased pollution). The rationale for this policy is that taxes on desirable activities create market distortions, discouraging increased levels of desirable actions. For example, economists argue that income taxes provide a number of distortions to the economy. "By taxing away some of the returns on working and saving, income taxes deter some people from joining the labor force and encourage others to consume too much of their income [instead of saving]."[139]

Using carbon tax revenues in this manner has been described as yielding a double-dividend: (1) reduced GHG emissions and (2) reduced market distortions from the taxes on desirable behavior. In the early 1990s, some economists suggested that the double-dividend effect would be strong enough to keep the overall costs of carbon reductions to society relatively small or even negative.[140] However, more recent economic studies indicate that the additional tax-imposed costs[141] would probably (at least in the United States)[142] exceed the economic efficiency benefits (unrelated to climate change benefits) from reducing the distortionary taxes.[143]

Direct or Indirect Payments to Households

To alleviate the carbon tax-imposed costs to households, particularly low-income households, policymakers could use the tax revenues to provide financial assistance. There are multiple mechanisms through which households could receive assistance, including lump-sum payments, payroll or income tax deductions, tax credits or rebates, or increases in existing assistance programs.

Provide Assitance to Carbon-Intensive Industries

Stakeholders in carbon-intensive industries—e.g., fossil fuel-fired electric utilities, steel producers, paper manufacturing, etc.—argue that if Congress enacts a GHG emission control program, such as a carbon tax, their industries would be adversely impacted. To alleviate potential impacts, policymakers may consider returning—directly or indirectly—a portion of the carbon tax revenues to affected industries. This would be analogous to providing a percentage of emission allowances (or auction revenues) at no cost to covered sources in a cap-and-trade program.[144]

Fund Specific Objectives

Congress may also consider using the tax revenues to provide funding to support a range of objectives related to climate change, including:

- Technology development and deployment: Promotion of emission mitigation technology is widely recognized as a vital step towards making substantial GHG emission reductions.[145]
- Energy efficiency: Improvements in energy efficiency could make considerable contributions in achieving GHG emission reductions. Although a carbon tax should stimulate energy efficiency by placing a price on carbon, Congress may consider using the tax revenues to provide additional incentives, particularly to encourage households to adopt conservation and other measures.
- Biological sequestration: Trees, plants, and soils sequester carbon, removing it from the earth's atmosphere. Tax revenues could be used to promote carbon sequestration efforts: e.g., forestry or agricultural activities.
- Adaptation efforts: Regardless of emission reduction efforts taken today, climatic changes are expected, due to the ongoing accumulation of GHGs in the atmosphere. Therefore, some contend that investment (e.g., tax revenues) should focus on preparing communities—both domestically and internationally—to adapt to the effects of a changing climate.
- Deficit reduction: A portion of the revenues could be set aside to offset projected federal tax revenue shortfalls.[146] In addition, Congress may consider using car on tax revenues to offset shortfalls beyond those that are related to the carbon tax.

Policy Considerations of Different Revenue Applications

Keeping in mind the potential amount of new carbon tax revenue, its use and distribution could have appreciable impacts. In general, Congress would face a zero-sum game when deciding how to apportion the tax revenues: One revenue use necessarily forgoes the opportunity to apply that level of revenue elsewhere. Analogous to emission allowance value allocation[147] in a cap-and-trade program, the distribution of tax revenue would present policymakers with a series of tradeoffs. The primary options involve (1) minimizing the overall costs of the carbon tax imposed on society; (2) alleviating the costs borne by subgroups in society and economic sectors; and (3) providing funding to support other policy objectives, such as technological development.

In the short-term, these options are mutually exclusive. However, in the long-term, advancements in low-carbon technology have the potential to reduce societal costs.

Overall Costs to Society

Overall costs may be referred to as macroeconomic costs and are often measured in terms of changes in projected gross domestic product (GDP) or another societal-scale metric—e.g., efficiency cost or welfare changes. Economic studies indicate that using carbon tax or auction revenues to offset reductions in distortionary taxes (labor, income) would be the most economically efficient use of the revenues and yield the greatest benefit to the economy overall. These studies show that if the revenues are used for other purposes, economic efficiency could suffer and the overall cost of the program would be higher.

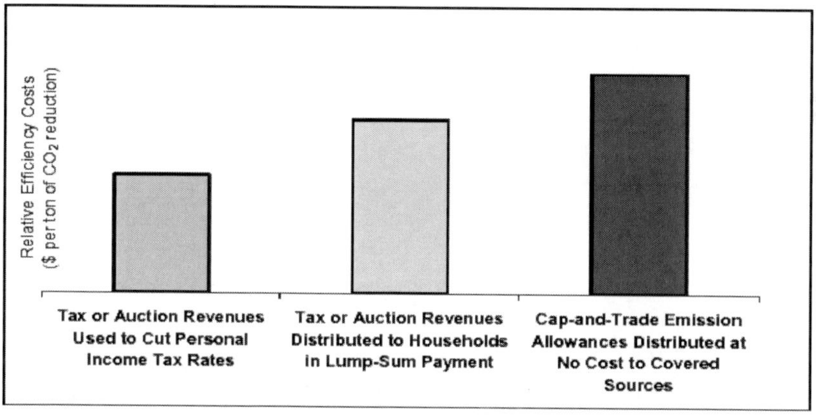

Source: Prepared by CRS with data from Lawrence H. Goulder, *Mitigating the Adverse Impacts of CO_2 Abatement Policies on Energy-Intensive Industries* (2002), Resources for the Future Discussion Paper.

Figure 9. Relative Differences in Efficiency Costs between Different Applications of Tax (or Auction) Revenues and No-Cost Allowance Distribution in a Cap-and-Trade Program.

A frequently cited study[148] compares the macroeconomic effects of different tax (or allowance auction) revenue applications with different emission allowance allocations in a cap-and-trade program. The study concludes that the least economically efficient method of emission allowance distribution would be to provide all of the emission allowances to covered sources at no cost. This is illustrated as the right-hand (red) column in **Figure 9**. This study finds that efficiency costs could be reduced by 21% from this scenario if policymakers were to distribute carbon tax revenues to U.S. households in "lump-sum" payments—e.g., increase the standard tax deduction or provide payments to households (represented by the middle/blue column in **Figure 9**). The lowest overall costs of those modeled would be achieved by using carbon tax revenues to reduce taxes on labor or investment. This option is shown in the left/green column. Compared to the least efficient option—no-cost distribution in a cap-and-trade program—this option would reduce societal costs by 47%. **Figure 9** indicates this option also produces substantially lower (approximately 32%) costs than lump-sum distribution. This is explained by the "double-dividend" effect, described above. However, the next section points out that the distributional impacts of these revenue applications are very different.

Distributional Impacts

In isolation, a carbon tax is regressive. This means that lower-income households would see a greater percentage increase of household income required for carbon tax-related costs—higher energy and/or electricity bills, primarily—than higher income households. A 2007 report estimated that households in the lowest income bracket could face increased costs equal to 3.4% of their household income; households in the highest bracket could see a cost increase of 0.8% (**Table 4**).[149] If Congress were to redistribute tax revenues to households in lump-sum payments, the 2007 report estimates the impacts to household income would be more progressive: The lowest income brackets would see net gains, while higher income brackets would break even or experience relatively minor losses (**Table 4**). In another scenario examined by the 2007 study, workers would receive a tax rebate equal to the payroll taxes paid—including worker and employer contributions—in a given year.[150] The rebate would be capped at the same level ($560/year) for each worker. As **Table 4** indicates, this option, while alleviating much of the regressiveness of the carbon tax, would disproportionately impact the two lowest income brackets.[151] This disproportionate outcome would be expected, because some of the lower income households may contain unemployed or retired persons, who would not benefit from a payroll tax rebate but still experience the higher energy prices brought on by the carbon tax. The last scenario in **Table 4**, in addition to the rebates for workers, would provide equal rebates to social security recipients. This addition would generate a more progressive distribution of tax-related impacts.

Table 4. Distributional Effects of Carbon Tax with Different Applications of Carbon Tax Revenues

Distributional Scenarios	Percentage Impact to Household Income by Income Bracket (1= Lowest, 10=Highest)									
	1	2	3	4	5	6	7	8	9	10
Carbon Tax in Isolation (before revenue recycling)	-3.4	-3.1	-2.4	-2.0	-1.8	-1.5	-1.4	-1.2	-1.1	-0.8
Carbon Tax with a Lump-Sum Disri-uion to Households	2.1	1.0	0.6	0.4	0.3	0.1	-0.1	-0.1	-0.2	-0.2
Carbon Tax with Payroll Tax Rebte for Workers	-0.7	-1.0	-0.2	0.1	0.1	0.3	0.2	0.2	0.0	0.0
Carbon Tax with Payroll Tax Rebate for Workers and Eqivalent Rebate for Social Secu-ity Recipients	1.4	1.0	0.6	0.3	0.1	0.1	0.1	-0.1	-0.1	-0.2

Source: Prepared by CRS with data from the following: Gilbert Metcalf, *A Proposal for a U.S. Carbon Tax Swap: An Equitable Tax Reform to Address Global Climate Change* (2007), The Hamilton Project, Brookings Institution.

Note: These results do not account for consumers' behavioral responses to the carbon tax. Metcalf points out that the results provide a "reasonable first approximation" of the different impacts to household incomes (Metcalf (2008)).

Carbon-Intensive Industries

This application of tax revenue would likely generate intense debate from industry and non-industry stakeholders alike. In general, stakeholders from all carbon-intensive industries argue that their businesses would be disproportionately affected by a carbon tax and thus should receive a portion of the revenues to avoid financial losses or provide transition into the new carbon-constrained regime. One counter-argument is that revenues used for this purpose would be unavailable to support other objectives (some of which are described in this report), but that concern applies to any use of the tax revenues. Regardless, many would assert that providing some level of assistance to carbon-intensive industries may be a political necessity.[152] Complicating the debate is the fact that the impacts of a carbon tax would vary across and within carbon-intensive industries. In the context of carbon tax impacts, there are two broad categories of carbon-intensive industries: industrial sectors that can more easily pass along tax-related costs and those that cannot. The first group would likely include fossil fuel producers and the electric utility sector. Fossil fuel producers and electricity generating facilities are generally expected to pass along the vast majority of their tax-related costs. Some contend that providing assistance to these industries would run counter to the principle of creating a level playing field for low-carbon energy sources. In addition, there are likely to be both winners and losers within the electricity generation sector. For instance, some facilities, such as coal-fired plants, are expected to see greater losses, while others—hydroelectric or renewable energy plants—may see financial gains under a carbon tax or cap-and-trade program.[153] The second category of carbon-intensive industries includes economic sectors that may face a comparative disadvantage if they were subject to GHG controls (directly or indirectly) and their international competition were not.[154] Because they may be constrained in their ability to pass along tax-related costs, there may be a stronger argument for providing assistance to this subset of carbon-intensive industries. However, it may be difficult for policymakers to determine the relative capacities of different industries and companies to pass along costs, or the degree of competitive disadvantage borne.[155] Policymakers have other options—e.g., to attach a carbon price to imported materials—that could address potential comparative disadvantages that may arise under a carbon tax system.[156] Recent legislative proposals would have provided some assistance to carbon-intensive industries. In the 110th Congress, Representative Larson's carbon tax proposal (H.R. 3416) would have allotted to "negatively affected" industries[157] approximately 8% of the tax revenues in the first year of the tax (2008 as written), declining to zero by 2018.[158] In addition, all of the cap-and-trade bills proposed in the 110th Congress that specified an emission allowance distribution strategy (as opposed to delegating the decision to a federal agency) would have allotted some percentage of allowance value to carbon-intensive industries.[159]

Technology Development

Many experts consider government investment in technology an essential component[160] of a robust strategy to mitigate GHG emissions, because long-term emission reduction would require major restructuring of the energy economy. Moreover, the development and deployment of lowcarbon technologies could have a considerable impact on the overall costs of an emissions reduction program. This is particularly the case if revenues induced technologies to be commercialized ahead of their projected schedules or if unanticipated low- or zero-carbon alternatives can be developed. Although dispensing tax revenues for technological development could yield considerable gain, the return in terms of additional investment to promote technology improvements remains uncertain.[161] For example, some programs may be operating at full capacity, and additional funding may not provide comparable benefits. Moreover, the marketplace already provides some incentive for technological change, and a carbon tax would increase the existing incentive to develop low-carbon or zero-carbon energy alternatives.[162] In addition, using federal funds to support technology raises a number of issues for policymakers that are beyond the scope of this report.[163]

CONCLUSION

Market-based mechanisms that limit greenhouse gas (GHG) emissions can be divided into two types: quantity control (e.g., cap-and-trade) and price control (e.g., carbon tax). To some extent, a carbon tax and cap-and-trade program would produce similar effects: Both would place a price on carbon, and both are estimated to increase the price of fossil fuels, which would ultimately be borne by energy users, particularly households. In an economically efficient market with perfect information, either a price (carbon tax) or quantity control instrument (cap-and-trade system) could be designed to achieve the same outcome. Because this market ideal does not exist, preference for a carbon tax or a cap-and-trade program ultimately depends on which variable one prefers to control—emissions or costs. Although there are several design mechanisms that could be included with either program that would blur the distinction, the gap between price control and quantity control can never be completely overcome. Compared to a cap-and-trade system, a carbon tax may both offer several advantages as well as present several possible disadvantages. A known price on carbon would not add to the volatility of energy prices, such as electricity and household fuels, unlike what may occur under a cap-and-trade program. Moreover, with a set price, industry would have better information to guide investment decisions (e.g., energy efficiency improvements and/or equipment upgrades), unless the tax rate was subject to periodic adjustment to achieve GHG emission reduction goals. Economists often highlight the relative economic efficiency advantage of a carbon tax, but this advantage rests on assumptions about the expected costs and benefits of climate change mitigation, both of which contain considerable uncertainty and some controversy. Some contend that a carbon tax may provide implementation advantages, including greater transparency, reduced administrative burden, and relative ease of modification. In addition, some argue that a carbon tax approach may make it easier to reach a GHG reduction agreement with developing nations. However, some counter that a cap-and-trade approach is best suited to meet this objective. The primary disadvantage of a carbon tax is that it would yield uncertain emissions. For some this concern may present a non-starter, precluding a carbon tax as an option to control GHG emissions. Some argue that the potential for irreversible climate change impacts necessitates the emissions certainty that is only available with a quantity-based instrument (e.g., cap-and-trade). However, depending upon its design, a cap-and-trade program may also include a risk of uncertain emissions. For example, if offsets are allowed as a compliance option, there is some concern that they would not represent real emission reductions. Emissions leakage offers another threat to the certainty of the emissions cap. Although it may present implementation challenges, policymakers could devise a tax program that yields only short-term emission fluctuations, as it progresses towards its long-term emission reduction objective. Proponents argue that short-term emission fluctuations would be preferable to price fluctuations that might be expected with a cap-and-trade system. Many assert that it would be easier to garner support for a cap-and-trade system, because a tax increase policy typically carries a political stigma. To some extent, this argument may be viewed by some as based on semantics, because a cap-and-trade program would act very much like a tax, in terms of its effects on energy prices. However, a carbon tax and a cap-and-trade legislation would fall into different committee jurisdictions within Congress. The committees developing the GHG emission legislation may have Members with varying perceptions about the relative values of uncertain costs versus uncertain emission reductions. The committee jurisdiction factor may play a role in the debate over the preferred policy instrument. In addition, some environmental groups may oppose a carbon tax, because they assume the tax rate would be set too low to be effective in controlling GHG emissions. This is more of an argument over stringency than over the instrument used to mitigate GHG emissions. The debate over a carbon tax versus a cap-and-trade program may be instructive for policymakers because it highlights (1) the primary differences between the instruments and the (2) design elements that could be included within either instrument that would blur their distinctions. For example, Congress could devise a cap-and-trade program that controls total costs and price volatility to some degree. However, a cap-and-trade price control mechanism, such as a safety-valve and/or reserve price, would alter the character of the cap-and-trade program—the level of emissions would no longer be certain. This potential outcome reinforces the fundamental choice between a carbon tax and a cap-and-trade program: cost certainty versus emissions certainty.

APPENDIX. ADDITIONAL INFORMATION

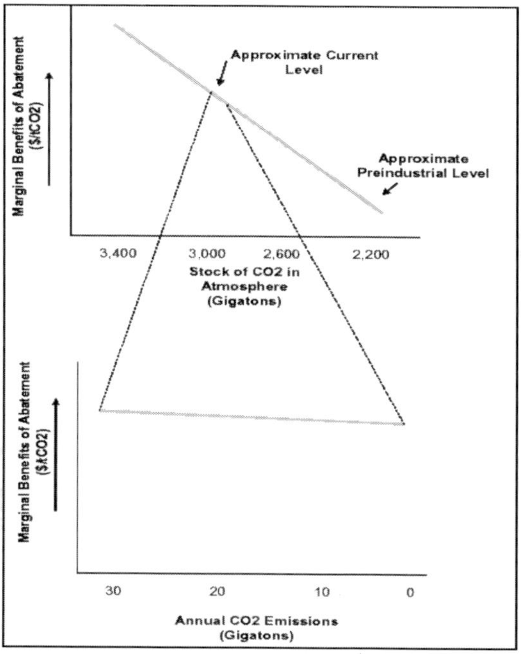

Source: Prepared by CRS. Adapted from William Pizer, *Prices vs. Quantities Revisited: The Case of Climate Change* (1997), Resources for the Future Discussion Paper 98-02.

Note: This figure illustrates the concept that for a stock pollutant, such as CO_2, reductions made in one year would have minimal effect on the pollutant's stock. As shown at the top of the figure, the current stock of CO_2 is approximately 3,000 gigatons (1 gigaton = 1 billion tons). Human-related activities generated approximately 27 gigatons of CO_2 emissions in 2005. If all nations were to stop emitting CO_2 emissions in one year, the effect on the stock of CO_2 would likely be minimal. Thus, the marginal benefit of abatement (for one year) is thought to be relatively flat, as depicted in the bottom of the figure. It would take multiple years of zero emissions to impact the current stock of CO_2 emissions.

Figure A-1. Illustration of Relationship between the Stock of CO_2 in Atmosphere and Annual CO_2 Emissions.

Table A-1. Comparison of Estimated Carbon Tax-Related Price Impacts to Fossil Fuels and Motor Gasoline from Selected Carbon Tax Rates

	Carbon Tax-Related Price Impacts			
Carbon Tax Rate	Coal (short ton)	Oil (barrel)	Natural Gas (thousand cubic feet, mcf)	Motor Gasoline (gallon)
\$5/$mtCO_2$	\$9.45/short ton	\$2.15/barrel	\$0.28/mcf	\$0.05/gallon
\$15/ $mtCO_2$	\$28.35/ short ton	\$6.45/ barrel	\$0.83/ mcf	\$0.14/ gallon
\$25/ $mtCO_2$	\$47.25/ short ton	\$10.75/ barrel	\$1.38/ mcf	\$0.23/ gallon
\$50/ $mtCO_2$	\$94.50/ short ton	\$21.50/ barrel	\$2.75/ mcf	\$0.45/ gallon
Inputs Used in Above Comparisona				
Metric tons of CO_2 $mtCO_2$) per unit of fuel[b]	1.89 $mtCO_2$/ short ton[c]	0.43 $mtCO_2$/ barrel[d]	0.055 tCO_2/ mcf[e]	0.009 tCO_2/ gallon[f]

a. These estimates assume that the carbon tax is applied upstream to fossil fuels. The price impacts assume that the carbon tax is passed through to end-users and not absorbed by upstream energy producers. The price effects also do not take into account behavioral changes—e.g., energy conservation, fuel substitution, etc.—in response to price increases.

b. Values derived from CO_2 coefficients (i.e., CO_2 emissions per quadrillion BTU) and thermal conversion factors (i.e., BTU per fuel unit)for each fuel. CO_2 coefficients are from EIA, *Documentation for Emissions of Greenhouse Gases in the United States 2006* (2008), table 6-1; thermal conversion factors from EIA, Annual Energy Review 2007 (2008), Appendix A.

c. This value represents the CO_2 coefficient for coal (electric power sector) and the thermal conversion factor for coal consumption from the electric power sector (2007).

d. This value represents the CO_2 coefficient for crude oil and the thermal conversion factor for "unfinished oil."

e. This value represents CO_2 coefficient for natural gas (pipeline) and the thermal conversion factor for enduse sectors (2007).

f. This value represents the CO_2 coefficient for "motor gasoline" and the thermal conversion factor for motor gasoline (conventional).

End Notes

[1] The major GHGs discussed include carbon dioxide (CO_2), methane (CH_4), nitrous oxide (N_2O), sulfur hexafluoride (SF_6), hydrofluorocarbons (HFC), and perfluorocarbons (PFC). Recent GHG reduction proposals have also included nitrogen trifluoride (NF_3).

[2] As discussed in this report, terminology is a key issue. Some proponents of the "carbon tax" approach describe the policy instrument as a user fee or user charge. There are multiple reasons that proponents may seek to change the nomenclature. Perhaps the primary concern is the political stigma associated with the word "tax." Regardless, the term "carbon tax" is the one that is most commonly associated with the GHG control policy instrument discussed in this report.

[3] Non-carbon GHGs could still be subject to the tax based on their contribution to global warming in relation to CO_2. Global warming potential (GWP) is an index of how much a GHG may contribute to global warming over a period of time, typically 100 years. GWPs are used to compare gases to carbon dioxide, which has a GWP of 1. For example, methane's GWP is 25, and is thus 25 times more potent a GHG than CO_2. The GWPs listed in this report are from: Intergovernmental Panel on Climate Change, *Climate Change 2007: The Physical Science Basis* (2007), p. 212.

[4] See e.g., Gilbert Metcalf, *A Green Employment Tax Swap: Using a Carbon Tax to Finance Payroll Tax Relief* (2007); Nathaniel Keohane and Sheila Olmstead, *Markets and the Environment* (2007), Island Press; Ian Parry "Fiscal Interactions and the Case for Carbon Taxes over Grandfathered Carbon Permits," in *Climate Change Policy* (Dieter Helm, editor), Oxford University Press (2005).

[5] Named after A. Cecil Pigou, author of a landmark economic work, *Economics of Welfare* (1920).

[6] See also, CRS Report RL33799, *Climate Change: Design Approaches for a Greenhouse Gas Reduction Program*, by Larry Parker; CRS Report RL34513, *Climate Change: Current Issues and Policy Tools*, by Jane A. Leggett.

[7] See e.g., William Pizer, *Prices vs. Quantities Revisited: The Case of Climate Change* (1997), Resources for the Future Discussion Paper 98-02.

[8] The marginal cost curve indicates the cost of an *additional unit* (e.g., one ton of GHG emissions) of emission abatement at different emission levels. A rising marginal cost curve (as depicted in Figure 1) signals that as each incremental unit of GHG emission abatement is made, the cost of abatement (per unit) increases.

[9] The marginal benefit curve indicates the benefit of an *additional unit* (e.g., one ton of GHG emissions) of emission abatement at different emission levels. A declining benefit slope (as depicted in Figure 1) illustrates that with each incremental unit of abatement, the per unit benefit decreases.

[10] Abatement may include emission reductions, emissions avoided, and sequestration activities.

[11] In other words, for abatement above this level, it would cost more than the value of the benefits received. Society would be paying too much; money would be better spent for other purposes.

[12] See CRS Report RL33970, *Greenhouse Gas Emission Drivers: Population, Economic Development and Growth, and Energy Use*, by John Blodgett and Larry Parker.

[13] For a analysis of some recent assessments, see CRS Report RL34489, *Climate Change: Costs and Benefits of S. 2191/S. 3036*, by Larry Parker and Brent D. Yacobucci.

[14] See also, CRS Report RL33799, *Climate Change: Design Approaches for a Greenhouse Gas Reduction Program*, by Larry Parker.

[15] Emission allowances are essentially the currency in a cap-and-trade program. An allowance is generally defined as a limited authorization by the government to emit 1 ton of pollutant. In the case of GHGs, an allowance generally refers to a metric ton of carbon dioxide equivalent (tCO_2-e).

[16] A 2008 Resources for the Future (RFF) report estimated the effects banking provisions had on price volatility. The authors concluded that banking eliminates about 20% of the cost differences between a price control (tax) system and cap-and-trade without banking provisions. Harrison Fell *et al.*, *Prices Versus Quantities Versus Bankable Quantities* (2008), Resources for the Future Discussion Paper.

[17] See CRS Report RL34502, *Emission Allowance Allocation in a Cap-and-Trade Program: Options and Considerations*, by Jonathan L. Ramseur.

[18] See Dieter Helm *et al.*, "Credible Carbon Policy," in *Climate Change Policy* (ed. Dieter Helm) (2005).

[19] M. L. Weitzman, "Price vs. Quantities" (1974), *Review of Economic Studies*, 41(4): 477-491.

[20] This conclusion is often referred to as the "Weitzman Rule." Nathaniel Keohane and Sheila Olmstead, *Markets and the Environment* (2007), Island Press

[21] In other words, not all emission reductions (or sequestration activities) are equal in terms of cost: there may be "low-hanging fruit" opportunities as well as actions that would require more substantial capital investments.

[22] Richard Newell and William Pizer, "Regulating Stock Externalities under Uncertainty," *Journal of Environmental Economics and Management* 45: 416-432 (2003); Warwick J. McKibbin and Peter J. Wilcoxen, "The Role of Economics in Climate Change Policy," (2002) *Journal of Economic Perspectives* 16(2): 107-29.

[23] The "curves" in this report's figures are depicted as straight lines for illustrative purposes.

[24] Although marginal benefit information is also uncertain, the relative-slope principle generally depends only on the marginal cost curve. The reasoning behind this approach and circumstances that may undermine this approach are discussed below.

[25] Other GHG emissions may yield damages unrelated to climate change impacts.

[26] 1 gigaton = 1 billion tons.

[27] William Pizer, *Prices vs. Quantities Revisited: The Case of Climate Change* (1997), Resources for the Future Discussion Paper 98-02.

[28] See e.g., James Hansen, Testimony before the House Select Committee on Energy Independence and Global Warming (June 20, 2008).

[29] See Timothy M. Lenton, *et al.*, "Tipping Elements in the Earth's Climate System," *Proceedings of the National Academy of Sciences* 105:6 (February 2008).

[30] Nicholas Stern, *The Economics of Climate Change: The Stern Review* (2006), p. 314 (emphasis in original).

[31] Nicholas Stern, *The Economics of Climate Change: The Stern Review* (2006), p. 314, footnote 12.

[32] William Pizer, *Climate Change Catastrophes* (2003), Resources for the Future Discussion Paper. See also, CBO, *Policy Options for Reducing CO_2 Emissions* (2008).

[33] See CRS Report RL34489, *Climate Change: Costs and Benefits of S. 2191/S. 3036*, by Larry Parker and Brent D. Yacobucci.

[34] CBO, *Policy Options for Reducing CO_2 Emissions* (2008).

[35] William Pizer, *Prices vs. Quantities Revisited: The Case of Climate Change* (1997), Resources for the Future Discussion Paper 98-02.

[36] Lawrence Goulder and Ian Parry, *Instrument Choice in Environmental Policy* (2008), Resources for the Future Discussion Paper, implicitly referencing: Michael Hoel and Larry Karp, "Taxes Versus Quotas for a Stock Pollutant," *Resource and Energy Economics* (2002) 24:367-84; Newell and Pizer (2003); and Larry Karp and Jiangfeng Zhang, "Regulation of Stock Externalities with Correlated Abatement Costs," *Environmental and Resource Economics* (2005) 32:273-299.

[37] Robert Stavins, "Correlated Uncertainty and Policy Instrument Choice," *Journal of Environmental Economics and Management* 30:218-232 (1996).

[38] Newell and Pizer (2003).

[39] Gilbert Metcalf *et al.*, *Analysis of U.S. Greenhouse Gas Tax Proposals* (2008), MIT Joint Program on the Science and Policy of Climate Change, Report No. 160.

[40] Pizer (1997).

[41] See, Hoel and Karp (2002); Newell and Pizer (2003); Karp and Zhang (2005).
[42] Ibid.
[43] Metcalf *et al.* (2008).
[44] Harrison Fell *et al.*, *Prices Versus Quantities Versus Bankable Quantities* (2008), Resource for the Future Discussion Paper.
[45] See Alan Ingham and Alistair Ulph, "Uncertainty and Climate Change Policy,"in *Climate Change Policy* (Dieter Helm, editor), Oxford University Press (2005).
[46] Indeed, the Carbon Tax Center lists price volatility control as one of the fundamental reasons taxes are superior to emissions caps. Carbon Tax Center, Tax vs. Cap-and-Trade, at [www.carbontax.org].
[47] William Nordhaus, "To Tax or Not to Tax: Alternative Approaches to Slowing Global Warming" (2007), *Review of Environmental Economics and Policy*, Vol. 1 (1): 26-44.
[48] Energy Information Administration, *Annual Energy Outlook 2008* (2008).
[49] Nordhaus (2007).
[50] Ian W.H. Parry and William Pizer, *Emissions Trading Versus CO_2 Taxes Versus Standards* (2007), Resources for the Future Issue Brief.
[51] See CRS Report RL34150, *Climate Change and the EU Emissions Trading Scheme (ETS): Kyoto and Beyond*, by Larry Parker.
[52] Although Phase 2 began in 2008, the market for Phase 2 allowances was created in 2005. Unlike Phase 1 allowances, Phase 2 allowances may be banked for future use in a planned Phase 3 anticipated to begin in 2013.
[53] See CRS Report RL34502, *Emission Allowance Allocation in a Cap-and-Trade Program: Options and Considerations*, by Jonathan L. Ramseur.
[54] Carbon Tax Center, "Tax vs. Cap-and-Trade," at http://www.carbontax.org/issues/carbon-taxes-vs-cap-and-trade/.
[55] See CRS Report RL34488, *Regulating a Carbon Market: Issues Raised By the European Carbon and U.S. Sulfur Dioxide Allowance Markets*, by Mark Jickling and Larry Parker.
[56] See CRS Report RL34436, *The Role of Offsets in a Greenhouse Gas Emissions Cap-and-Trade Program: Benefits and Concerns*, by Jonathan L. Ramseur.
[57] See CRS Report RL34489, *Climate Change: Costs and Benefits of S. 2191/S. 3036*, by Larry Parker and Brent D. Yacobucci.
[58] Metcalf (2008).
[59] AEI, 2007.
[60] Parry and Pizer, 2007.
[61] As with other comparisons, this would depend upon the design of the cap-and-trade program.
[62] See Kenneth Green, *et al., Climate Change: Caps vs. Taxes* (2007), American Enterprise Institute.
[63] *Ibid.*
[64] This potential problem could be alleviated by allowing covered sources to bank emission allowances.
[65] A cap-and-trade program, depending upon its design, may carry some risk of uncertain emissions as well. For example, if offsets are allowed as a compliance option, there is some concern that they would not represent real emission reductions. Emissions leakage is another threat to the certainty of the emissions cap. See CRS Report RL34436, *The Role of Offsets in a Greenhouse Gas Emissions Cap-and-Trade Program: Potential Benefits and Concerns*, by Jonathan L. Ramseur; and CRS Report R40100, *"Carbon Leakage" and Trade: Issues and Approaches*, by Larry Parker and John Blodgett.
[66] See e.g., Metcalf (2008) and Shapiro (2008).
[67] See Dieter Helm *et al.*, "Credible Carbon Policy," in *Climate Change Policy* (ed. Dieter Helm) (2005).
[68] Council on Foreign Relations, *Confronting Climate Change: A Strategy for U.S. Foreign Policy* (2008), Independent Task Force Report No. 161.

[69] See Letter from Representatives Sensenbrenner and Cantor to Representatives Conyers and Nadler (February 11, 2009), at http://www.eenews.net/features/documents/2009/02/11/document_pm_03.pdf. This letter also argues that if a cap-and-trade bill were to raise revenue, the bill should originate in the House of Representatives.

[70] See e.g., Gilbert Metcalf, *A Green Employment Tax Swap: Using a Carbon Tax to Finance Payroll Tax Relief* (2007).

[71] For example, the U.S. Climate Action Partnership (USCAP), a group of large companies (and environmental organizations) recommends that a "significant portion of allowances should be initially distributed free to capped entities and to economic sectors particularly disadvantaged by the secondary price effects of a cap.... " U.S. CAP, *A Call to Action*, at http://www.us-cap.org/.

[72] See CRS Report RL34150, *Climate Change and the EU Emissions Trading Scheme (ETS): Kyoto and Beyond*, by Larry Parker.

[73] See IPA Energy Consulting, *Implications of the EU Emissions Trading Scheme for the UK Power Generation Sector* (2005), Prepared for the United Kingdom Department of Trade and Industry; Jos Sijm et al., "CO2 Cost Pass-Through and Windfall Profits in the Power Sector," *Climate Policy* 6 (2006): 49-72; Point Carbon Advisory Services, *EU ETS Phase II—The Potential and Scale of Windfall Profits in the Power Sector* (2008), Prepared for World Wildlife Fund.

[74] See CRS Report RL33846, *Greenhouse Gas Reduction: Cap-and-Trade Bills in the 110th Congress*, by Larry Parker, Brent D. Yacobucci, and Jonathan L. Ramseur.

[75] For example, the President and CEO of Duke Energy, who initially was an advocate a cap-and-trade approach (and remains a member of USCAP), described the Lieberman-Warner bill (S. 2191) as a "bastardized" version of cap-andtrade due to its allocation scheme. See Jim Rogers, "Climate Change Legislation Should Not Be Punitive" (February 2008), *Energy Daily*.

[76] Hepburn (2006), citing P.D. Klemperer, *Auctions: Theory and Practice* (2004), Princeton University Press.

[77] In a January 2009 speech, the CEO of ExxonMobil (Rex Tillerson) spoke in favor of a carbon tax over a cap-andtrade system, stating a carbon tax would be "the most efficient means of reflecting the cost of carbon in all economic decisions—from investments made by companies to fuel their requirements, to the product choices made by consumers." This quotation was taken from a UPI news report, at [www.upi.com].

[78] There may be some industry stakeholders who support a carbon tax on the assumption that it will be politically challenging for Congress to subsequently raise carbon taxes to the level needed to stimulate the stringent emissions reductions targeted (e.g., 80% below current levels by 2050) in recent cap-and-trade proposals.

[79] For example, different treatment of offsets could raise concern, particularly if covered entities in the United States gain access to questionable offset projects. See CRS Report RL34436, *The Role of Offsets in a Greenhouse Gas Emissions Cap-and-Trade Program: Potential Benefits and Concerns*, by Jonathan L. Ramseur.

[80] Qualified credits could range from EU emission allowances (EUAs) from the EU ETS to certified emission reductions (CERs)—i.e., international offsets—issued by the Clean Development Mechanism of the Kyoto Protocol.

[81] Scott Barrett and Robert Stavins, "Increasing Participation and Compliance in International Climate Change Agreements," *International Environmental Agreements: Politics, Law and Economics*, Vol. 3:349-76 (2003).

[82] However, Kyoto participants account for the majority of 1990 emissions (the Kyoto baseline).

[83] Data from World Resources Institute, Climate Analysis Indicators Tool (CAIT), at http://cait.wri.org/.

[84] See CRS Report RL34150, *Climate Change and the EU Emissions Trading Scheme (ETS): Kyoto and Beyond*, by Larry Parker; and CRS Report RL33826, *Climate Change: The*

Kyoto Protocol, Bali "Action Plan," and International Actions, by Susan R. Fletcher and Larry Parker.

[85] William Nordhaus and Joseph Boyer, *Warming the World: Economic Modeling of Global Warming* (2000), MIT Press.

[86] William Nordhaus, "To Tax or Not to Tax: Alternative Approaches to Slowing Global Warming" (2007), *Review of Environmental Economics and Policy*, Vol. 1 (1): 26-44.

[87] David Victor, *The Collapse of the Kyoto Protocol and the Struggle to Slow Global Warming* (2001), Princeton University Press.

[88] EPA, *Tools of the Trade: A Guide to Designing and Operating a Cap and Trade Program for Pollution Control* (2003), p. 3-5.

[89] For additional reading on these considerations, see e.g., Gilbert Metcalf and David Weisbach, *The Design of a Carbon Tax* (June 2008), Tufts University and the University of Chicago; EPA, *Tools of the Trade* (2003).

[90] Non-CO_2 GHG emissions come from a wide range of sources, most of which are outside the energy sector.

[91] Power plants may be considered midstream, because their emissions are influenced by the demand from electricity consumers: businesses and households. In addition, some carbon-intensive industries, such as steel or cement, might be categorized as midstream.

[92] Section 821 of the 1990 Clean Air Act Amendments requires electric generating facilities affected by t he acid rain provisions of Title IV to monitor CO_2 in accordance with Environmental Protection Agency (EPA) regulations.

[93] Although a price signal would be sent in either case—thus (in theory) encouraging CCS development—some contend that if coal is taxed upstream at the extraction point, some of the price signal may be weakened before it reaches coal-fired emission sources.

[94] For more information on this technology, see CRS Report RL33801, *Carbon Capture and Sequestration (CCS)*, by Peter Folger.

[95] See e.g., John Reilly *et al.*, "The Role of Non-CO_2 GHGs in Climate Policy: Analysis Using the MIT IGSM" (2006), *The Energy Journal*, Special Issue No. 3: 503-520.

[96] Metcalf *et al.* (2008). The overall costs fall, because the tax rate is decreased. By including all GHGs, the tax rate can be lowered if the cumulative target remains the same.

[97] Gilbert Metcalf and David Weisbach, *The Design of a Carbon Tax* (June 2008), Tufts University and the University of Chicago.

[98] There are additional GHGs, but they are not included in the current U.S. GHG inventory. For more information, see CRS Report RL34266, *Climate Change: Science Highlights*, by Jane A. Leggett.

[99] Based on 2006 data. EPA, *Inventory of U.S. Greenhouse Gas Emissions and Sinks: 1990-2006* (April 2008).

[100] For example, a 5-cent-per-barrel tax on the oil industry—domestically produced oil and oil imports—supports the Oil Spill Liability Trust Fund.

[101] In 2006, there were 3,292 plants that had at least one fossil fuel-fired generator (EIA, 860 data files).

[102] EIA, *Annual Coal Report 2006* (2007).

[103] Industry used 5% of the coal in 2006; residences and commercial entities used 3%.

[104] EPA, *Inventory of U.S. Greenhouse Gas Emissions and Sinks: 1990-2006* (April 2008).

[105] Metcalf and Weisbach (2008).

[106] Some of the cap-and-trade proposals made in recent years would delegate coverage decisions to the implementing agency, typically EPA.

[107] Alternatively, policymakers could develop emission/sequestration standards for specific activities. This latter option would not be a market-based approach, but it was included in one of the cap-and-trade proposals from the 110th Congress—H.R. 6186 (Markey).

[108] See EPA's Landfill Methane Outreach Program, at http://www.epa.gov/lmop/overview.htm.

[109] A small amount of perfluorocarbons (PFCs) are also used as substitutes (EPA Inventory, 2008). Under the Montreal Protocol and the Clean Air Act Amendments of 1990, ODSs are

being phased out. See CRS Report RL30853, *Clean Air Act: A Summary of the Act and Its Major Requirements*, by James E. McCarthy et al.

[110] EPA Inventory (2008).

[111] See e.g., Metcalf and Weisbach (2008).

[112] Intergovernmental Panel on Climate Change, *Safeguarding the Ozone Layer and the Global Climate System, Issues related to Hydrofluorocarbons and Perfluorocarbons* (2005).

[113] As with fossil fuels, policymakers may want to consider applying the tax to imported steel and cement. This would increase the number of entities subject to the tax.

[114] See e.g., Metcalf and Weisbach (2008).

[115] EPA Inventory (2008).

[116] EPA Inventory (2008).

[117] A comprehensive discussion and analysis of climate change costs and benefits is beyond the scope of this report. This section highlights issues associated with estimating and applying climate change cost and benefit estimates for carbon tax implementation.

[118] IPCC, *Climate Change 2007: Mitigation*, (Annex I—Glossary)Contribution of Working Group III to the Fourth Assessment Report of the Intergovernmental Panel on Climate Change (2007).

[119] Richard Tol, "The Social Cost of Carbon: Trends, Outliers, and Catastrophes," *Economics E-Journal* (August 2008). This updates a previous study: Richard Tol, "The Marginal Damage Costs of Carbon Dioxide Emissions: An Assessment of the Uncertainties," *Energy Policy* 33:2064-2074 (2005)

[120] See e.g., Nicholas Stern, *The Economics of Climate Change: The Stern Review* (2006); Richard Tol, "The Marginal Damage Costs of Carbon Dioxide Emissions: An Assessment of the Uncertainties," *Energy Policy* 33:2064-2074 (2005); David Pearce, "The Social Cost of Carbon," in *Climate Change Policy* (Dieter Helm, editor), Oxford University Press (2005).

[121] For a more comprehensive discussion and explanation of discount rates and their applicability to climate change, see David Weisbach and Cass Sunstein, *Climate Change and Discounting the Future: A Guide for the Perplexed*, Reg-Markets Center Working Paper No. 08-19 (August 2008); and Jiehan Guo *et al.*, "Discounting and the Social Cost of Carbon: A Closer Look at Uncertainty" *Environmental Science and Policy*, 9: 205-126 (2006).

[122] Intergovernmental Panel on Climate Change, *Climate Change 2007: Impacts, Adaptation, and Vulnerability* (2007), Working Group II Contribution to the Fourth Assessment Report, p. 823.

[123] Nicholas Stern, *The Economics of Climate Change: The Stern Review* (2006), p. 26.

[124] Stern's business-as-usual estimate for the SCC was $95 per metric ton of CO_2-e (in $2005); the SCC estimate for a pathway that would stabilize GHG concentration at 450 parts per million was $28.

[125] Ibid., p. 301.

[126] Gilbert Metcalf, *A Proposal for a U.S. Carbon Tax Swap: An Equitable Tax Reform to Address Global Climate Change* (2007), The Hamilton Project, Brookings Institution.

[127] Nicholas Stern, *The Economics of Climate Change: The Stern Review* (2006), p. 301.

[128] See Robert Shapiro et al., *Addressing Climate Change Without Impairing the U.S. Economy: The Economics and Environmental Science of Combining a Carbon-Based Tax and Tax Relief* (2008), U.S. Climate Task Force; and Gilbert Metcalf, *A Proposal for a U.S. Carbon Tax Swap: An Equitable Tax Reform to Address Global Climate Change* (2007), The Hamilton Project, Brookings Institution.

[129] Richard Tol, "The Social Cost of Carbon: Trends, Outliers, and Catastrophes," *Economics E-Journal* (August 2008).

[130] If non-peer-reviewed estimates were included, the average would be much higher, because these estimates are generally in the upper end of the range. See Richard Tol, "The Social Cost of Carbon: Trends, Outliers, and Catastrophes," *Economics E-Journal* (August 2008).

[131] Intergovernmental Panel on Climate Change, *Climate Change 2007: Impacts, Adaptation, and Vulnerability* (2007), Working Group II Contribution to the Fourth Assessment Report.

[132] For example, William Nordhaus estimates a SCC of approximately \$8/tCO_2-e (*A Question of Balance: Economic Modeling of Global Warming* (2008), prepublication version,Yale Press); Nicholas Stern's SCC estimate is approximately \$26/tCO_2-e. Both of these estimates are in \$2005 and are for different business-as-usual scenarios (Nicholas Stern, *The Economics of Climate Change: The Stern Review* (2006)).

[133] This is the projected emission reduction from the Lieberman-Warner proposal (S. 2191/S. 3036).

[134] Selecting a GHG reduction goal would be a complicated endeavor, involving multiple considerations. This discussion is beyond the scope of this report.

[135] For a comprehensive analysis of these issues see CRS Report RL34489, *Climate Change: Costs and Benefits of S. 2191/S. 3036*, by Larry Parker and Brent D. Yacobucci.

[136] Although some (primarily economists) may define this phrase more narrowly—i.e., only in the context of offsetting distortionary taxes—in this report, revenue recycling refers to any application of carbon tax revenue.

[137] Neither bill would address other GHG emissions.

[138] U.S. Internal Revenue Service, *Internal Revenue Service Data Book 2005* (2006)..

[139] Ian W.H. Parry and William Pizer, *Emissions Trading Versus CO_2 Taxes Versus Standards* (2007), Resources for the Future Issue Brief.

[140] Ian Parry, "Fiscal Interactions and the Case for Carbon Taxes over Grandfathered Carbon Permits," in *Climate Change Policy* (Dieter Helm, editor), Oxford University Press (2005); Intergovernmental Panel on Climate Change Working Group III, *Climate Change 2001: Mitigation*, (Cambridge, UK: Cambridge University Press, 2001) Chapter 8.

[141] This is referred to as the "tax-interaction effect" in economic literature. In short, a carbon tax would increase energy costs. Increased energy costs are expected to lead to some decrease in real household income and labor supply. See e.g., Ian Parry, "Fiscal Interactions and the Case for Carbon Taxes over Grandfathered Carbon Permits," in *Climate Change Policy* (Dieter Helm, editor), Oxford University Press (2005)

[142] There is some evidence that the double-dividend benefits are stronger in Europe than in the United States, because the former has a more stringent income tax system. With more stringent taxes in place, there is more "distortion" to reduce, thus strengthening the double-dividend effect. Intergovernmental Panel on Climate Change Working Group III, *Climate Change 2001: Mitigation*, (Cambridge, UK: Cambridge University Press, 2001) Chapter 8, citing several studies.

[143] For cost estimates from economic models, see Metcalf *et al.* (2008) and Robert Shapiro *et al.*, *Addressing Climate Change Without Impairing the U.S. Economy: The Economics and Environmental Science of Combining a Carbon-Based Tax and Tax Relief* (2008), U.S. Climate Task Force.

[144] For more discussion, see CRS Report R40100, *"Carbon Leakage" and Trade: Issues and Approaches*, by Larry Parker and John Blodgett.

[145] See e.g., CBO, *Evaluating the Role of Prices and R&D in Reducing Carbon Dioxide Emissions*, (2006).

[146] A carbon tax would raise energy costs, which is expected to lead to some reduction in real household wages and labor supply. In turn, federal tax revenue may decline.

[147] Allowance value allocation includes both the use of auction revenues and distribution of allowances to entities (e.g., industry or non-covered parties) at no cost.

[148] In this study, the author applies a carbon tax of approximately \$7/tCO_2 that rises (at 7% or 9%) until it reaches \$14/tCO_2, at which point the tax is held constant in real dollars. The study also examines an emissions trading program with allowance prices that match the carbon tax paths. Lawrence H. Goulder, *Mitigating the Adverse Impacts of CO_2 Abatement Policies on Energy-Intensive Industries* (2002), Resources for the Future Discussion Paper.

[149] Gilbert Metcalf, *A Proposal for a U.S. Carbon Tax Swap: An Equitable Tax Reform to Address Global Climate Change* (2007), The Hamilton Project, Brookings Institution.

[150] The Social Security payroll tax applies a constant tax rate—half of which is paid by the employee and the other half paid by the employer (the self-employed are responsible for the entire amount)—up to a certain income level. According to the Tax Policy Center, for nearly two-thirds of U.S. households, payroll taxes (if the employer portion is counted) are the single largest tax paid to the federal government at: http://www.taxpolicy center.org/publications/url.cfm?ID=1001065.

[151] Because this option effectively replaces a portion of the payroll tax (a regressive tax on an economic good) with a carbon tax (a levy on an economic bad), it has the potential to improve economic efficiency, and thus lower overall society costs. However, Metcalf asserts that the rebate is unlikely to appreciably alter labor supply, except among lowincome workers. See Metcalf (2007).

[152] For a further discussion of rationales for providing support to carbon-intensive industries, see CRS Report R40100, *"Carbon Leakage" and Trade: Issues and Approaches*, by Larry Parker and John Blodgett.

[153] Dallas Burtraw and Karen Palmer, *Compensation Rules for Climate Policy in the Electricity Sector* (2007), Resources for the Future Discussion Paper.

[154] Among others, the steel, paper, and aluminum industries are often cited as examples of trade-exposed industries.

[155] Department of Climate Change, Commonwealth of Australia, *Carbon Pollution Reduction Scheme: Green Paper* (July 2008), p. 310.

[156] See CRS Report R40100, *"Carbon Leakage" and Trade: Issues and Approaches*, by Larry Parker and John Blodgett.

[157] As determined by the Secretary of the Treasury in consultation with the Secretary of Labor.

[158] The other carbon tax proposal from the 110th Congress—H.R. 2069 (Stark)—does not include revenue distribution provisions.

[159] See CRS Report RL33846, *Greenhouse Gas Reduction: Cap-and-Trade Bills in the 110th Congress*, by Larry Parker, Brent D. Yacobucci, and Jonathan L. Ramseur.

[160] Some argue that the private sector is likely to under-invest in research and development, even with a carbon price in place. See W. David Montgomery and Anne E. Smith, *Price, Quantity, and Technology Strategies for Climate Change Policy* (2005), CRA International.

[161] See Robert W. Hahn, *Greenhouse Gas Auctions and Taxes: Some Practical Considerations* (2008), AEI Center for Regulatory and Market Studies.

[162] See CRS Report RL34621, *Capturing CO2 from Coal-Fired Power Plants: Challenges for a Comprehensive Strategy*, by Larry Parker, Deborah D. Stine, and Peter Folger.

[163] See CRS Report RL34513, *Climate Change: Current Issues and Policy Tools*, by Jane A. Leggett.

In: Carbon Taxes
Editor: Neil Viveiros

ISBN: 978-1-62618-148-9

Chapter 3

CARBON TAXES: A REVIEW OF EXPERIENCE AND POLICY DESIGN CONSIDERATIONS*

Jenny Sumner, Lori Bird and Hillary Smith

EXECUTIVE SUMMARY

State and local governments in the United States are evaluating a wide range of policies to reduce carbon emissions, including, in some instances, carbon taxes. Carbon taxes have existed internationally for nearly 20 years.

This report reviews existing carbon tax policies both internationally and in the United States. It also analyzes carbon policy design and effectiveness. Table ES-1 summarizes the carbon tax rates, annual revenue collected, and revenue distribution methods of existing and proposed carbon tax policies.

While each carbon tax is designed to meet the needs of the implementing jurisdiction, the implementation of all carbon taxes involves design considerations such as which sectors to tax, where to set the tax rate, how to use tax revenues, what the impact will be on consumers, and how to ensure emissions reduction goals are achieved.

Carbon taxes are most commonly placed on gasoline, coal, and natural gas. There are large differences among tax rates across various jurisdictions. The highest rates are paid in Sweden ($105 per metric ton CO_2), while the

* This is an edited, reformatted and augmented version of the National Renewable Energy Laboratory, Technical Report NREL/TP-6A2-47312, dated December 2009.

lowest rates are paid in the Bay Area Air Quality Management District (BAAQMD) in California ($0.045 per metric ton CO_2). How jurisdictions set tax rates depends on what outcomes they are trying to achieve; higher rates can induce behavior change, while lower rates may be more appropriate to generate funds for carbon mitigation programs. Some jurisdictions, such as Sweden, allow industries to pay a reduced tax rate.

Table ES-1. Overview of Carbon Tax Policies

Country/ Jurisdiction	Start Date	Tax Rate ($USD unless noted otherwise)	Annual Revenue	Revenue Distribution
Finland	1990	$30/metric ton CO_2 (€20)	$750 million (€500 million)	Government budget; accompanied by independent cuts in income taxes
Netherlands	1990	~$20/metric ton CO_2 in 1996	$4.819 billion[a] (€3.213 billion)	Reductions in other taxes; Climate mitigation programs
Norway	1991	$15.93 to $61.76/metric ton CO_2 (NOK 89 to NOK 345)	$900 million (1994 estimate)	Government budget
Sweden	1991	Standard rate: $104.83/metric ton CO_2 (910 SEK) Industry rate: ~$23.04/metric ton CO_2 (~200 SEK)	$3.665 billion (25 billion SEK)	Government budget
Denmark	1992	$16.41/metric ton CO_2 (90 DKK)	$905 million	Environmental subsidies and returned to industry
United Kingdom	2001	$0.0078/kWh for electricity; $0.0027/kWh for natural gas provided by gas utility; $0.0175/kg for liquefied petroleum gas or other gaseous hydrocarbons supplied in a liquid state; and $0.0213/kg for solid fuel	$1.191 billion (£714 million)	Reductions in other taxes
Boulder, CO	2007	$12-13 per metric ton CO_2	$846,885	Climate mitigation programs
Quebec	2007	$3.20 per metric ton of CO_2 (C$3.50)	$191 million (C$200 million)	Climate mitigation programs
British Columbia	2008	$9.55 per metric ton of CO_2 in 2008 (C$10), increasing $4.77 (C$5) annually to $28.64 (C$30) in 2012	$292 million (C$306 million)	Reductions in other taxes

Country/ Jurisdiction	Start Date	Tax Rate ($USD unless noted otherwise)	Annual Revenue	Revenue Distribution
BAAQMD, California	2008	$0.045 per metric ton of CO_2e[b]	$1.1 million (expected)	Climate mitigation programs
France	proposed	$24.74 per metric ton of CO_2 (€17)	$4.499 billion (€3 billion) expected	Reductions in other taxes
CARB, California	proposed	$0.155 per metric ton CO_2e in FY 2010-11, dropping to $0.09 per metric ton CO_2e in 2014	$63.1 million 2010-2013; $36.2 million starting in 2014, expected	Climate mitigation programs

[a] Revenue in the Netherlands is from all environmentally related taxes, of which carbon taxes are the clear majority.

[b] CO_2e is carbon dioxide equivalent.

Revenue from carbon taxes can be distributed in several ways. Some jurisdictions (e.g., Boulder, Quebec, BAAQMD) use the revenue for carbon mitigation programs. Other jurisdictions (e.g., United Kingdom, British Columbia, France's proposed policy) return carbon tax revenue to customers. Sweden and Norway use carbon tax revenues to fund general government budgets. Finland's carbon tax also is directed to the general government budget, but it is accompanied by a cut in income tax. In some cases, carbon tax revenues have been redirected to low-income consumers to minimize impacts.

While carbon taxes are often touted as the more economically efficient means of addressing carbon emissions, they may not ensure a set level of emissions reductions. Carbon taxes can be designed so that tax rates increase automatically if emission reductions are not met, though this practice has not been fully implemented to date.

Emission reductions that are due to carbon taxes can be difficult to measure. Some jurisdictions have quantified reductions in overall emissions, while other jurisdictions have examined impacts that are due to programs funded by carbon tax revenues.

Carbon tax policies can also be integrated with other carbon policies to maximize greenhouse gas (GHG) reductions. In Europe, countries implementing a carbon tax also operate within the European Union's Emissions Trading Scheme. Other complementary policies such as Norway's tax on hydrofluorocarbons (HFCs) and perfluorocompounds (PFCs) can address different GHGs.

1. Introduction

Carbon taxes have existed internationally for nearly 20 years. In the early 1990s, carbon taxes emerged in northern European countries, with Finland being the first nation to adopt a carbon tax in 1990. Subsequently, the Netherlands (1990), Norway (1991), Sweden (1991), and Denmark (1992) implemented carbon taxes. More than a decade after the first carbon tax, in 2001, the United Kingdom began its Climate Change Levy.

Interest in federal carbon policy has recently increased in the United States. The U.S. House of Representatives passed H.R.2454 "The American Clean Energy and Security Act of 2009" ("Waxman-Markey bill") in June 2009, and the U.S. Senate introduced S.1733 "Clean Energy Jobs and American Power Act" ("Kerry-Boxer bill") in September 2009. While recent federal policy has focused on carbon cap and trade policies, Congress introduced three carbon tax bills in 2009.[1] Proposals vary, but each taxes fossil fuel production and imports between $10 and $15 per short ton of carbon dioxide (CO_2) in the first year and creates plans to increase the tax over time.

As a policy mechanism, carbon taxes have been extensively compared to carbon cap and trade systems. In terms of economic efficiency, carbon taxes are expected to generate more net benefits than a cap and trade system, which is inherently inflexible. The Congressional Budget Office (CBO) surveyed current literature and estimated that the net benefits of a tax could be roughly five times those of an inflexible cap with no banking, borrowing, or other cost containment mechanisms (CBO 2008). One benefit of a carbon tax is that it continuously encourages emissions reductions, whereas a cap and trade system only encourages reductions to the point of the cap. Further, taxes are straightforward to administer and less subject to gaming. Taxes are also viewed as beneficial because the price is known, in contrast with a cap and trade system, which can create volatile prices, because of the unpredictable nature of supply, demand, and regulatory conditions (Nordhaus 2008). Also, taxes create revenue, which can be used to address distributional inequities by returning it to low-income consumers.

However, the primary criticism of taxes is that they do not necessarily ensure a certain level of emissions reductions as emissions caps do. Carbon taxes are often viewed as less politically acceptable than cap and trade systems. Despite the distinctions often made between the two policies, a pure carbon cap and trade system could be modified to address benefits that carbon taxes provide. A carbon cap and trade system could allow banking and borrowing of allowances or other cost containment mechanisms, thus decreasing price volatility. It could also raise revenue—by auctioning allowances—and distribute the revenue to low-income consumers. A review of the first compliance period of the European Union's Emission Trading Scheme (ETS) found that the program evolved "surprisingly well" and was able to establish a transparent price on tradable CO_2 emission allowances (Ellerman and Joskow 2008).

In the most recent wave of carbon taxes, state and local governments have become more active in adopting taxes as either a complement to federal policies or a reaction to lack of federal leadership. In 2007, Boulder, Colorado and Quebec, Canada introduced carbon taxes, and in 2008, British Columbia, Canada, and the Bay Area Air Quality Management District (BAAQMD) in California began implementing carbon taxes. France and the California Air Resources Board (CARB) proposed carbon taxes in 2009.

There is also increasing interest in combining carbon taxes with other carbon mitigation policies. Carbon cap and trade systems may not cover all sectors or may be more difficult to implement in sectors without point sources (e.g., transportation); therefore, carbon taxes may be designed to work in conjunction with other carbon policies. For example, the recently proposed French carbon tax is designed to cover sectors not addressed by the European Union's ETS.

This paper discusses considerations in the design of carbon taxes and reviews experience implementing carbon taxes domestically and internationally. It also discusses the effectiveness of these policies and difficulties measuring impacts. Measures of effectiveness can vary by jurisdiction and specific policy goals. While all carbon taxes at their core are designed to reduce CO_2 or GHG emissions, some taxes may be aggressive in their efforts to encourage changes in behavior and consumption patterns, while others may be implemented at more modest levels to generate revenues to support the expansion or cost effectiveness of low-carbon technologies. Effectiveness has been evaluated based on (1) aggregate carbon emissions reductions in the jurisdiction, (2) emissions reductions that are due to the tax, and (3) emissions reductions from programs implemented with tax revenue.

2. Carbon Tax Design Considerations

Carbon taxes place a value on CO_2 and other GHG emissions, thus internalizing some portion of the costs associated with their environmental impact. While all carbon taxes inherently provide this function, policy goals may vary. Carbon taxes serve primarily to reduce GHG emissions by placing a cost on emissions, but can also raise revenues to provide funding for carbon mitigation programs or create market signals for consumers. Policy design considerations associated with implementing carbon taxes include determining the tax base, which sectors to tax, where to set the tax rate, how to use tax revenues, how to assess the impact on consumers, and how to ensure the tax achieves emissions reduction goals.

Tax Base

To implement carbon taxes, governments must decide which fuels or sources to place the tax. Most commonly, carbon taxes are placed on gasoline, coal, and natural gas. Some governments, however, exempt certain industries from carbon taxes or allow those industries to pay lower tax rates.

Governments must also decide whether to place the tax on upstream or downstream sources of emissions. Taxing upstream sources may provide an administratively efficient method of tax collection, while taxing downstream sources such as electricity consumption may provide a more direct signal to consumers. Although this debate is not the focus of this paper, it is important to recognize the potential effects of taxing upstream versus downstream sources when designing a carbon tax.

When considering which sectors to tax, some governments may also be limited by their jurisdictions. For example, the BAAQMD has authority to control air pollution for non-vehicular stationary sources, thus the primary sectors taxed include petroleum refineries, power plants, cement plants, and landfills. Table 1 summarizes the major fuels or sources on which current carbon taxes are applied in the 12 jurisdictions described in Section 3.

Tax Rate

As shown in Figure 1, carbon tax rates vary across jurisdictions, in part due to their function. Higher carbon tax rates provide stronger signals to consumers to change behavior, while lower rates may not do much to change behavior but can provide funds for carbon mitigation programs.

Some of the highest rates are seen in Europe[2]: Sweden's standard tax rate is the equivalent of $105 per metric ton CO_2; however, the rate for industry is substantially less at $23 per metric ton CO_2. Norway's tax on gasoline equates to \$62 per metric ton CO_2 and Finland's tax is \$30 per metric ton CO_2. France's proposed tax rate is modeled after the current price for CO_2 allowances in the European Union's ETS and is set at the equivalent of about \$25 per metric ton CO_2.

Some of the lowest rates occur in California. The BAAQMD rate is \$0.045 per metric ton CO_2, and the BAAQMD explicitly designed the tax to raise revenue to support local GHG mitigation programs rather than to encourage behavior change. The proposed CARB tax rate is also set comparatively low, at \$0.155 per metric ton CO_2, and is designed to generate funds for GHG mitigation programs.

Revenue Distribution

Revenues from carbon taxes are directed in different ways. Revenues can be (1) directed specifically to carbon mitigation programs, (2) directed to individuals through measures, such as reductions in income taxes, or (3) used to supplement government budgets. The choice of revenue distribution could impact the political sustainability of the tax.

Table 1. Major Taxed Sectors in Existing and Proposed Carbon Tax Systems

	Finland	**Netherlands**	**Norway**	**Sweden**	**Denmark**	**United Kingdom**	**France**	**Quebec**	**British Columbia**	**Boulder, CO**	**BAAQMD, CA**	**CARB**
Natural gas	√	√	√	√	√	√	√	√	√			
Gasoline	√		√	√			√	√	√			
Coal	√			√	√	√	√	√				
Electricity	√	√			√	√		√		√		
Diesel	√		√					√	√			
Light and heavy fuel oil	√	Light only	√	√	√			√				
Liquified petroleum gas (LPG)			√	√		√		√				
Home heating oil		√		√				√				
Permitted facilities											√	√

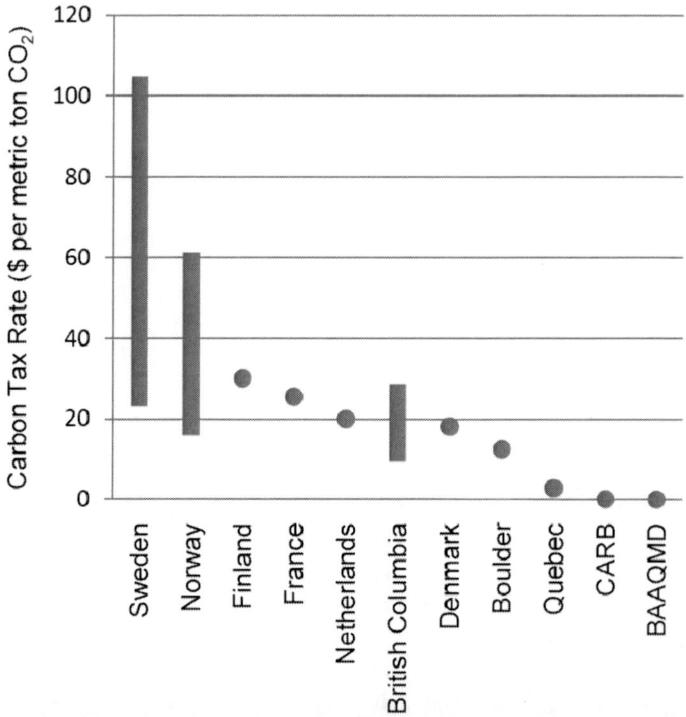

Notes: Sweden and Norway's carbon tax rates vary depending on the taxed sector. British Columbia's carbon tax was C$10 ($9.55) per metric ton CO_2 in 2008 but is scheduled to increase by C$5 per year to C$30 ($28.64) in 2012.

Figure 1. Carbon tax rates.

State and local governments frequently direct carbon tax revenues to carbon reduction programs that they implement. Boulder, Quebec, and the BAAQMD all use revenues for carbon mitigation programs, and CARB has proposed to do so as well.

Some carbon tax programs return tax revenue to customers through other means such as income tax reductions. The United Kingdom and British Columbia use this method, and France has proposed refunding revenue as well. Finland's approach does not earmark the carbon tax revenue, but the tax is accompanied by independent cuts in income taxes (Parkkinen 2009). These "revenue-neutral" mechanisms, which are designed to change customer behavior while reducing other taxes, do not raise money for government general funds or emission reduction programs. The "double dividend" theory suggests that revenue-neutral policies create two benefits: a price is placed on goods that harm the environment and lowered income taxes will spur new employment opportunities (OECD 2001). A revenue-neutral approach can also lower the overall economic impacts of implementing a carbon tax (Repetto 2001). Cost savings are also greater when tax revenues are returned through cuts in distorting taxes (such as income taxes) than when they are returned in lump sums. These results have been both demonstrated by economic theory and supported by numerical simulations. (Goulder 1995)

On the other hand, some carbon taxes, like those in Sweden and Norway, are used specifically to raise revenue for governments. Directing revenues to general government budgets can be easier to administer than funding carbon mitigation programs. However, critics say that carbon taxes simply are a way to raise government revenue rather than provide environmental benefits. If governments were to raise tax rates purely to increase revenue, the resulting tax would not be economically efficient—for an efficient tax, the rate should be set equal to the marginal damage caused by carbon emissions. An alternative approach would direct revenue to programs that address environmental impact through earmarking, thus limiting the incentive to raise taxes purely to generate more revenue. (Prasad 2008)

Impact on Consumers

When designing a carbon tax, the impact on low-income households is also a consideration; a common criticism of carbon taxes is that they disproportionately burden low-income households. Several policies, including income tax reductions and credits to low-income households, can be used to mitigate this concern. For example, British Columbia provides a low-income "climate action tax credit," a reduction of 5% in the first two personal income tax rates, and has proposed providing a Northern and Rural Homeowner benefit of up to $200 in 2011 (Ministry of Finance, British Columbia 2008). British Columbia estimated the net impact of the carbon tax and other associated tax cuts for two respective family types (seniors and a two-earner family of four), finding that the tax cuts exceed the cost of the carbon tax in both cases.[3] (Ministry of Finance, British Columbia 2008)

In France, President Nicolas Sarkozy proposed plans to return all revenues to households and businesses through reductions in income taxes or through issuing a "green check" (Keller 2009). While not explicitly addressing low-income consumers, France's proposal would help low- income households adapt to the carbon tax.

Carbon taxes impact businesses. Businesses may prefer carbon taxes to other carbon mitigation policies, because taxes provide a certain, long-term price signal that can be incorporated into projections of operating expenses, whereas prices may not be as well known with emissions caps. Energy-intensive industries or highly competitive industries that compete with companies in jurisdictions without taxes have expressed concerns about carbon taxes. To address concerns about the impact of carbon taxes on businesses, some jurisdictions allow certain industries to pay reduced rates. Sweden's reduced rate for industry is approximately \$23 per metric ton CO_2, while the standard rate is \$105 per metric ton CO_2. In Denmark, businesses that sign an energy savings agreement with the Ministry of Transportation and Energy can pay reduced rates. Denmark also reduced existing taxes on energy when it established a carbon tax, so that the effective tax rate was the same.

Ensuring Emissions Reductions

One of the key arguments against carbon taxes and in favor of emissions limits is that taxes do not necessarily guarantee emissions reductions. While a tax policy could be structured so that rates automatically increase if emissions reduction goals are not met, this approach has not been implemented. This may reflect the political challenges of passing a tax in the first place. In fact, many taxes are structured in ways to make them more politically feasible. Revenue-neutral policies and refunds to low-income consumers are designed to make policies more politically appealing.

Some governments have increased their tax rates over time, but none has implemented a policy to increase automatically if emissions reduction targets are not met. British Columbia's carbon tax is set to phase in over a four-year period and the government has indicated it will modify the tax if necessary to achieve carbon emissions goals over time; however, because the tax was only recently implemented, there have not been any attempts to modify the tax rate yet. Governments operating GHG-reduction programs may find it easier to modify programs than to raise taxes. Quebec publishes an annual progress report on its GHG-reduction programs. The City of Boulder increased rates to the maximum allowed by their ordinance and directed the additional revenues to implement more GHG-reduction programs. The BAAQMD has indicated that it may raise fees in the future, depending on the need of its Carbon Protection Program.

3. Carbon Taxes Implemented since 1990

While the first implemented carbon tax is approaching its twentieth year of implementation, there have been few reviews of carbon tax policies. Current literature focuses on the economic efficiency of carbon taxes compared to carbon cap and trade systems. This section examines existing carbon taxes, specifically looking at the sectors taxed, the tax rate, and annual revenue generated.

Table 2. Overview of Carbon Tax Policies

Country/ Jurisdiction	Start Date	Tax Rate ($USD unless noted otherwise)	Annual Revenue	Revenue Distribution
Finland	1990	\$30/metric ton CO_2 (€20)	\$750 million (€500 million)	Government budget; accompanied by independent cuts in income taxes
Netherlands	1990	~\$20/metric ton CO_2 in 1996	\$4.819 billion[a] (€3.213 billion)	Reductions in other taxes; Climate mitigation programs
Norway	1991	\$15.93 to \$61.76/metric ton CO_2 (NOK 89 to NOK 345)	\$900 million (1994 estimate)	Government budget
Sweden	1991	Standard rate: \$104.83/metric ton CO_2 (910 SEK) Industry rate: ~\$23.04/metric ton CO_2 (~200 SEK)	\$3.665 billion (25 billion SEK)	Government budget
Denmark	1992	\$16.41/metric ton CO_2 (90 DKK)	\$905 million	Environmental subsidies and returned to industry
United Kingdom	2001	\$0.0078/kWh for electricity; \$0.0027/kWh for natural gas provided by gas utility; \$0.0175/kg for liquefied petroleum gas or other gaseous hydrocarbons supplied in a liquid state; and \$0.0213/kg for solid fuel	\$1.191 billion (£714 million)	Reductions in other taxes
Boulder, CO	2007	\$12-13 per metric ton CO_2	\$846,885	Climate mitigation programs
Quebec	2007	\$3.20 per metric ton of CO_2 (C\$3.50)	\$191 million (C\$200 million)	Climate mitigation programs

Table 2. (Continued)

Country/ Jurisdiction	Start Date	Tax Rate ($USD unless noted otherwise)	Annual Revenue	Revenue Distribution
British Columbia	2008	$9.55 per metric ton of CO_2 in 2008 (C$10), increasing $4.77 (C$5) annually to $28.64 (C$30) in 2012	$292 million (C$306 million)	Reductions in other taxes
BAAQMD, California	2008	$0.045 per metric ton of CO_2e[b]	$1.1 million (expected)	Climate mitigation programs
France	proposed	$24.74 per metric ton of CO_2 (€17)	$4.499 billion (€3 billion) expected	Reductions in other taxes
CARB, California	proposed	$0.155 per metric ton CO_2e in FY 2010-11, dropping to $0.09 per metric ton CO_2e in 2014	$63.1 million 2010-2013; $36.2 million starting in 2014, expected	Climate mitigation programs

[a] Revenue in the Netherlands is from all environmentally related taxes, of which carbon taxes are the clear majority.

[b] CO_2e is carbon dioxide equivalent.

Finland—1990

Finland was the first country to adopt a carbon tax. Launched in 1990, Finland's carbon tax is a separate component of Finland's excise tax on fossil fuels used for transportation or heating. The carbon tax applies to gasoline, diesel, light fuel and heavy fuel oil, jet fuel, aviation gasoline, coal, and natural gas (Ministry of the Environment, Finland 2008a). Coal is subject to a tax of $73.97 (€49.32) per metric ton, natural gas is subject to a reduced tax rate of $3.02 (€2.016) per MWh, and liquid fuels are taxed between $0.07 (€0.05) and $0.09 (€0.06) per liter (European Environment Agency 2009). Commercial vessels and commercial air traffic as well as fuels used for electricity are exempt. Electricity is taxed, but the rate per kWh does not vary according to carbon content; however, a refund is available for renewable electricity. (Parkkinen 2009)

Initially, Finland based its carbon tax purely on carbon content but modified it to include a 60% carbon component and a 40% energy component. The energy component was a tax based on energy use in MWh rather than on carbon content of the fuel. In January 1997, however, Finland returned to a pure carbon tax (Parkkinen 2009). The carbon tax was most recently increased by 13% to $30 (€20) per metric ton CO_2 on January 1, 2008 (Ministry of Environment Finland 2008b). Carbon tax revenues have been approximately $750 million (€500 million) annually (Ministry of Environment Finland 2008b). All revenue from the carbon tax in Finland goes directly into the general central government budget without any earmarking (Parkkinen 2009).

In 2000, the Finnish government determined that because of the carbon tax, CO_2 emissions were reduced by roughly 4 million metric tons of CO_2 between 1990 and 1998. Four million metric tons of CO_2 represented approximately 7% of the 57 million metric tons of emissions recorded in 1998 (Prime Minister's Office, Finland 2000).

The Netherlands—1990

The Netherlands' carbon tax began in 1990. It applies to natural gas, electricity, blast furnaces, coke ovens, refinery and coal gas, coal gasification gas, gasoline, diesel, and light fuel. In 1996, the tax rate was equivalent to $20 per metric ton CO_2.

Environmentally related taxes generate total revenue of $4.819 billion (€3.213 billion), of which the carbon tax is the majority (Netherlands Ministry of Housing, Spatial Planning and the Environment n.d.).

The Netherlands uses carbon tax revenues to reduce the general tax burden for individuals and businesses as well as to provide programs to reduce greenhouse gases. Part of the revenue is "recycled" to businesses in the form of accelerated depreciation for environmental equipment and tax-deductibility of energy investments (Netherlands Ministry of Housing, Spatial Planning and the Environment, Directorate-General for Environmental Protection, n.d.).

The Netherlands Ministry of Housing, Spatial Planning and the Environment (n.d.) estimated the carbon tax would reduce annual CO_2 emissions 1.7 to 2.7 million metric tons annually in 2000, with covered sectors reducing CO_2 emissions by approximately 5%. Due to an increase in the tax in 1999, annual CO_2 emissions are projected to be reduced 3.6 to 3.8 million tons by 2010 and 4.6 to 5.1 million tons in 2020.

Norway—1991

Norway's carbon tax began in 1991. The current tax rates vary from $15.93 (NOK 89) per metric ton CO_2 to $61.76 (NOK 345) per metric ton CO_2

(Ministry of the Environment, Norway 2007). Taxed sectors include gasoline, light and heavy fuel oil, and oil and gas in the North Sea. The pulp and paper industry, fishmeal industry, domestic aviation, domestic shipping of goods, and the continental shelf (supply fleet) pay reduced rates. Foreign shipping, fishing in Norway, fishing in distant waters, and external aviation are all exempt from the tax. Table 3 shows Norwegian carbon tax rates by taxed sector in 2007. The tax is estimated to cover approximately 68% of Norway's CO_2 emissions, or approximately 50% of Norway's GHG emissions (UNFCCC 2006).

Norway directs carbon tax revenues to general government accounts. With this revenue and revenue from offshore drilling licenses, Norway has financed a special pension fund that contained $373 billion, or nearly $80,000 for every Norwegian, at the end of 2007 (Turner 2008).

Norway has received negative press because its GHG emissions increased by 15% from when it implemented a carbon tax in 1991 to 2008. Others justify the increase in GHG emissions by noting that Norway also experienced an increase in gross domestic product of 70% since 1990. (Abboud 2008)

Table 3. Norwegian Carbon Tax Rates, 2007

	Tax rate – US$ (NOK) per liter oil and petrol, kg. coal and coke or Sm^3 gas)[a]	Tax rate - NOK ($) per metric ton CO_2
Petrol (gasoline)	$0.14 (0.80 NOK)	$61.76 (345 NOK)
Fuel oil		
Light fuel oil, diesel	$0.10 (0.54 NOK)	$36.34 (203 NOK)
Heavy fuel oils	$0.10 (0.54 NOK)	$30.79 (172 NOK)
Mineral oils, lower rate		
Light fuel oil, diesel	$0.05 (0.28 NOK)	$18.80 (105 NOK)
Heavy fuel oils	$0.05 (0.28 NOK)	$15.93 (89 NOK)
Use of gas in Norway		
Natural gas	$0.08 (0.47 NOK)	$35.98 (201 NOK)
LPG	$0.11 (0.60 NOK)	$35.80 (200 NOK)
Continental shelf		
Light fuel oil, diesel	$0.14 (0.80 NOK)	$53.70 (300 NOK)
Heavy fuel oils	$0.14 (0.80 NOK)	$45.65 (255 NOK)
Natural gas	$0.14 (0.80 NOK)	$61.22 (342 NOK)

[a] Sm^3 refers to a standard cubic meter.

Industrial efficiency gains in Norway may have been achieved because of the carbon tax. Emissions per unit of production were 22% lower in 2003 than they were in 1991, but since 1996, emissions per unit of production have been steady or even rising slightly. Norwegian officials recognize that the carbon tax is not the only factor contributing to changes in industrial efficiency. Other factors that impact industrial efficiency include general technology improvements and the use of more mature petroleum fields, which provide increased energy per unit of production. (Ministry of the Environment, Norway 2005a)

Norway's carbon tax has promoted technology innovation in the form of carbon sequestration. In 1996, StatoilHydro, the company operating the Sleipner gas field, began developing a CO_2-sequestration project, storing CO_2 under the sea floor. The project, which cost approximately $200 million, has saved the company approximately $60 million in carbon taxes each year and has stored approximately 1 million metric tons of CO_2 annually since 1996 (UNFCCC 2006, Abboud 2008). However, even though the company has increased the efficiency of its operations, its total emissions have increased because of expansions in drilling (Abboud 2008).

Sweden—1991

In 1991, Sweden initiated a carbon tax of $44.37 (250 SEK) per metric ton CO_2. Industries, including manufacturing, agriculture, co-generation plants, forestry and aquaculture, pay a lower proportion of the general level (Ministry of Sustainable Development, Sweden 2005a). As shown in Figure 2, Sweden modified rates such that industry paid only $11.28 (80 SEK) per metric ton while other consumers paid $45.15 (320 SEK) per metric ton in 1993. The standard rate was $55.57 (370 SEK) per metric ton in 1996, and between 1999 and 2003, the standard rate rose to $104.83 (910 SEK) per metric ton while the rate for industry leveled off at approximately $23.04 (200 SEK) per metric ton.

Revenue from Sweden's carbon tax was relatively steady from 1993 to 2000 then gradually increased from 2000 to 2004 (Figure 3). It generated approximately $3.65 billion (25 billion SEK) annually in 2005, 2006, and 2007 (Swedish Tax Agency 2008a).[4] Sweden directs the tax revenues to the general government budget.[5]

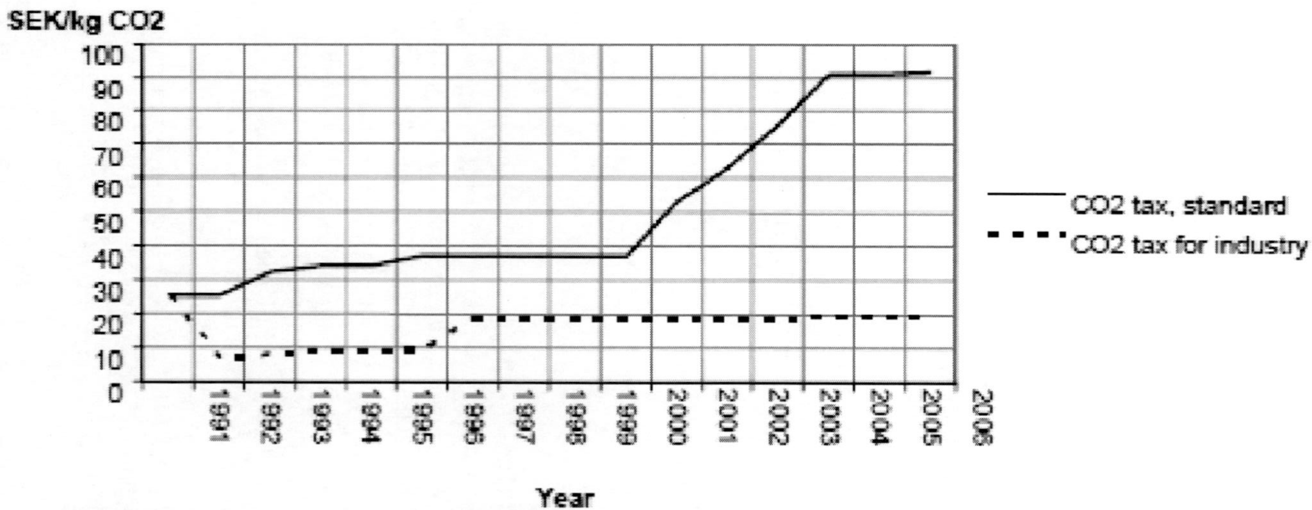

Source: Swedish Environmental Agency and Swedish Energy Agency, 2007.

Figure 2. Swedish CO_2 tax rate, SEK per kilogram CO_2, nominal figures.

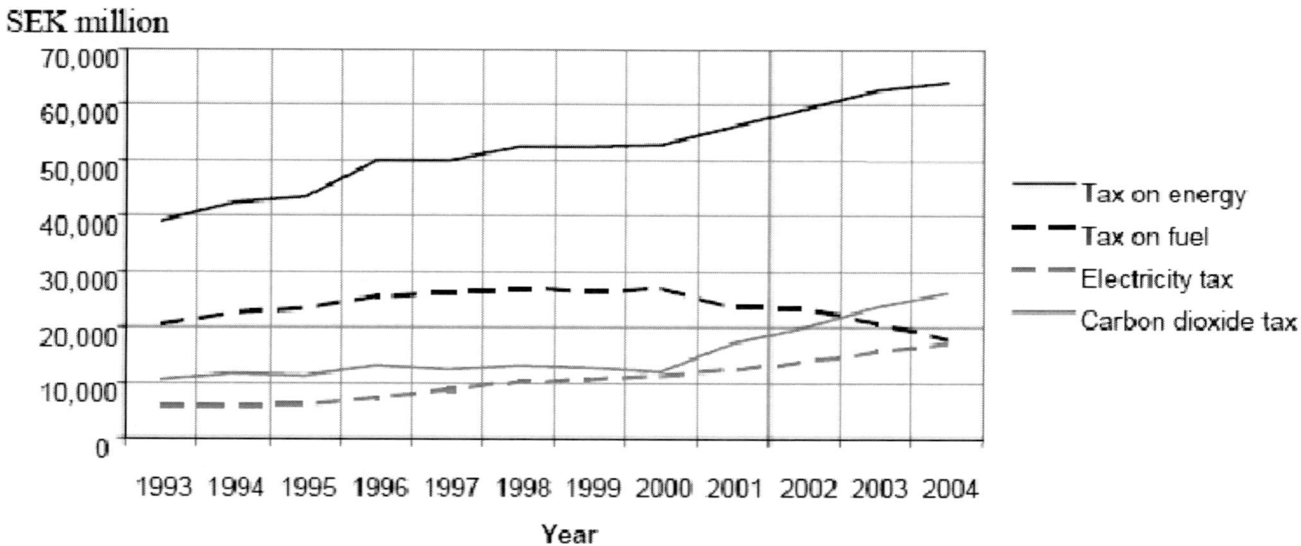

Source: Swedish Environmental Agency and Swedish Energy Agency, 2007.

Figure 3. Swedish carbon tax revenues.

Several studies in Sweden have attempted to assess CO_2 emissions. The Swedish Ministry of the Environment estimated that CO_2 emissions fell about 15% between 1995 and 1990 because of the tax (Johansson 2000). In terms of overall emissions, Sweden saw GHG emissions fall by almost 9% between 1990 and 2006. In December 2008, Sweden reported that GHG emissions have fallen by more than 40% since the mid-1970s. (Ministry of the Environment, Sweden 2008)

Denmark—1992

Denmark's tax on CO_2 passed in 1991 and took effect in May 1992. Fossil fuels are subject to both an energy tax and a CO_2 tax. When the CO_2 tax passed, it included a subsequent decrease in the energy tax to maintain the overall tax rate. Table 4 shows Danish energy and CO2 tax on various fuel types. The initial tax rate of $16.91 (100 DKK) per metric ton CO_2 was reduced to $16.41 (90 DKK) per metric ton CO_2 in 2005. With the decrease in the CO_2 tax in 2005, the energy tax increased to maintain the overall tax rate (Speck et al. 2006). In 2008, the carbon tax rate remained at $16.41 (90 DKK) per metric ton CO_2 emissions (Ministry of Taxation, Denmark 2008). In 2008, the revenues from the CO_2 tax were approximately $905 million.

In Denmark, approximately 40% of tax revenue is used for environmental subsidies, while the other 60% is returned to industry. The government provides an approximately 25% reduction of the tax if a company signs an energy savings agreement with the Ministry of Transportation and Energy. Per capita emissions in Denmark were reduced by 15% from 1990 to 2005 (Prassad 2008). Industrial emissions were also estimated to decrease by 23% during the 1990s, after adjusting for growth and market-induced industry restructuring (Enevoldsen 2005).

United Kingdom—2001

The UK Climate Change Levy (CCL) began in 2001. It imposes a tax on electricity, natural gas supplied by a gas utility, liquefied petroleum gas or other gaseous hydrocarbons supplied in a liquid state for heating, and solid fuel (e.g., coal and coke, lignite, semi-coke of coal or lignite, and petroleum coke). CCL rates only apply to industrial and commercial energy supplies to the industrial, commercial, agricultural, public and service sectors (UK

Department of Energy and Climate Change 2009). The CCL was designed to encourage businesses to become more energy efficient and reduce GHG emissions. Table 5 shows the average CCL rates as of April 1, 2009 by fuel type.

Table 4. Danish Energy and CO_2 Tax on Various Fuel Types[a]

		1985	1990	1996	2000	2002	2005
Light fuel oil (EURcent/l)	Energy tax	4.61	22.4	20.25	23.21	24.63	25
	CO_2 tax			3.67	3.62	3.63	3.23
	TOTAL TAX	4.61	22.4	23.92	26.83	28.26	28.23
Heavy fuel oil (EURcent/kg)	Energy tax	5.11	25.2	22.56	26.16	27.72	28.09
	CO_2 tax			4.35	4.29	4.31	3.9
	TOTAL TAX	5.11	25.2	26.9	30.45	32.03	31.99
Natural gas (EURcent/nm^3)[b]	Energy tax			0.14	21.47	27.19	27.42
	CO_2 tax			2.99	2.95	2.96	2.69
	TOTAL TAX			3.13	24.42	30.15	30.11
Pit coal (EURcent/kg)	Energy tax	1.62	9.8	11.69	17.44	19.25	19.49
	CO_2 tax			3.26	3.22	3.23	2.96
	TOTAL TAX	1.62	9.8	14.95	20.66	22.47	22.45

Source: Speck et al. 2006

[a] Tax rates expressed in national currencies have not been increased since 2002 with the result that current tax rates in real terms are slightly lower than they were in 2002.

[b] The unit nm^3 represents a normal cubic meter.

Table 5. United Kingdom Climate Change Levy Rates as of April 1, 2009

Sector	Tax rate
Electricity	\$0.0078 (£0.00470) per kWh
Natural gas as supplied by a gas utility	\$0.0027 (£0.00164) per kWh
Liquified petroleum gas (LPG) or other gaseous hydrocarbons supplied in a liquid state, for heating	\$0.0175 (£0.0105) per kg [\$0.0012 (£0.00076) per kWh]
Solid fuel, e.g., coal and coke, lignite; semi-coke of coal or lignite; and petroleum coke	\$0.0213 (£0.0128) per kg [\$0.0027 (£0.00164) per kWh]

Source: Hatherall, 2009.

CCL revenues in 2006-2007 were \$1.191 billion (£714 million), far less than the approximately \$2.25 billion (£1.35 billion) required to offset a 0.3%

cut in National Insurance Contributions[6] that was implemented with the CCL to make the policy revenue-neutral.[7] As a result, industry as a whole has not faced higher taxes; instead, the tax burden as a whole has been reduced because the cut in National Insurance Contributions is greater than the carbon tax (UK Department of Energy and Climate Change 2009). The CCL was estimated to generate approximately $1.167 billion (£700 million) in 2007-2008 and in 2008-2009 (Her Majesty's Treasury 2008).

A study by Cambridge Econometrics (2005) estimated that by 2010 the CCL would reduce energy demand by approximately 15% (12.8 million metric tons) in the commercial and public sectors (Her Majesty's Treasury 2008:101).

Boulder, Colorado—2007

Boulder, Colorado, a city of approximately 100,000 people located northwest of Denver, adopted a carbon tax via a ballot initiative in 2006. The tax, collected by the investor-owned utility, took effect in April 2007.

Rates are set based on kilowatt-hour (kWh) use of electricity, with residents initially paying $0.0022 per kWh, commercial customers paying $0.0004 per kWh, and industrial customers paying $0.0002 per kWh (City of Boulder 2009a). In August 2009, the tax increased to the maximum rate allowed by the tax ordinance: $0.0049 per kWh for residents, $0.0009 per kWh for commercial customers and $0.0003 per kWh for industrial customers (City of Boulder 2009a). The rates are equivalent to approximately $12-13 per ton of CO_2. The ordinance exempts the portion of electricity that Boulder residents can voluntarily purchase as utility-provided wind power through Xcel Energy (Boulder Revised Code 3-12-2).

Tax revenues were $846,885 in 2008 and are expected to peak at $1,609,000 in 2010 then decrease to $1,203,000 in 2012 as fossil fuel use decreases.[8] The tax is set to expire on March 31, 2013. Extension of the tax would require voter approval.

The City of Boulder uses its carbon tax revenue to fund a climate action plan that promotes energy efficiency in homes and buildings, renewable energy, and reductions in vehicle miles traveled. Boulder developed a set of guiding principles to evaluate its carbon reduction programs. Each program must both maximize GHG reductions and be cost effective. The city measures cost effectiveness in tax dollars spent per GHG ton reduced. Programs must also include a reasonable expectation for private investment and payback time. (City of Boulder 2009a) Boulder's goal is to reduce GHG emissions to a level that is 7% below the level of 1990 (1,478,534 metric tons CO_2e[9]) by 2012,[10] and it measures the success of the carbon tax in terms of meeting this goal. Boulder estimated emission reductions because of its climate action strategies as follows: (City of Boulder 2009b)

- Renewable energy activities: approximately 60,000 metric tons CO_2e in 2008
- Transportation: 33,000 metric tons CO_2e annually
- Energy efficiency programs: approximately 6,700 metric tons CO_2e in 2008.

Quebec, Canada—2007

Quebec implemented its carbon levy in October 2007. The carbon levy is $3.20 (C$3.50) per metric ton of CO_2. The level is readjusted every year based on the volume of sales.[11] The rate varies for each fuel, with the gasoline levy at $0.0076 (C$0.008) per liter, diesel fuel at $0.0086 (C$0.009) per liter, propane at $0.0048 (C$0.005) per liter, and coal at $7.64 (C$8.00) per metric ton.

According to Torys Climate Change Bulletin (2007), the carbon levy is expected to raise $191 million (C$200 million) annually as follows:

- $65.87 million (C$69 million) from gasoline producers
- $76.37 million (C$80 million) from diesel and heating oil producers
- $41 million (C$43 million) from electricity and natural gas producers
- $6.68 million (C$7 million) from coal and propane producers.

Quebec deposits its carbon tax revenue into a "green fund," which supports reductions in GHG emissions and improvements to public transportation. GHG-reduction measures—with the largest projected reduction in or avoidance of GHG—include (Quebec 2008):

- Requiring manufacturers of light-duty vehicles to meet a GHG-emissions standard starting in 2010
- Improving the energy efficiency in the transport of merchandise
- Financially supporting the capture and incineration or valorization of landfill gas
- Supporting manure processing and biomass
- Supporting research and innovation for carbon sequestration.

Quebec's 2006-2012 carbon mitigation action plan, *Quebec and Climate Change—A Challenge for the Future* (Quebec 2008), recommends 26 actions to reduce GHG emissions and adapt to climate change. The plan also is supported by C$350 million from the federal trust fund for clean air and climate change. In total, the plan is expected to reduce GHG emissions by 14.6 million metric tons by 2012 (Quebec 2008). GHG-emission reductions from programs funded by the tax are estimated at 11.2 million metric tons by 2012.[12] If GHG emissions held steady at 2005 levels of 92 million metric tons, this would represent a decrease of approximately 12%.

The government publishes an annual progress report, which enables it to refocus program priorities if necessary to meet its goals (Quebec 2008). It is unclear whether the government may choose to raise the carbon tax as part of this review process.

Quebec quantified GHG-reduction measures in terms of their reduction potential and total costs for the period 2006-2012. The following policies and programs are expected to be most cost effective (Quebec 2008):

- Blending a minimum of 5% ethanol in fuel by 2012
- Reducing industrial emissions
- Financially supporting landfill gas incineration or use
- Supporting manure processing and energy use from biomass.

British Columbia, Canada—2008

British Columbia, Canada launched its carbon tax in July 2008. The tax is applied primarily to transportation fuels, natural gas, and fuels used in industrial processes. It began at a level of \$9.55 (C\$10) per metric ton CO_2 and increases \$4.77 (C\$5) per metric ton CO_2 annually until reaching a level of \$28.64 (C\$30) per metric ton of CO_2 in 2012.

In 2008, the carbon tax was estimated to be an additional \$0.0223 (C\$0.0234) per liter of gasoline. The tax will increase to approximately \$0.0637 (C\$0.0667) per liter in 2012. Natural gas is taxed at \$0.47 (C\$0.50) per GJ, and coke is taxed at \$23.74 (C\$24.87) per metric ton (European Environment Agency 2009). After being phased in, the government may adjust the tax rate depending on whether its GHG emissions targets are being met, the impacts other carbon policies are having, the actions other governments are taking, as well as general advice being offered by its Climate Action Team (Ministry of Finance, British Columbia 2008).

The tax base includes fuel used to generate heat for households and industrial processes such as producing cement and drying coal. In addition, the carbon tax applies to road, rail marine, and air transportation within British Columbia. Inter-jurisdictional transportation is exempt from the carbon tax.

Carbon tax revenues for the 2008-2009 fiscal year were $292 million (C$306 million). The tax is expected to raise $531 million (C$557 million) in the 2009-2010 fiscal year, $714 million (C$748 million) in the 2010-2011 fiscal year and $911 million ($955 million) in the 2011-2012 fiscal year.[13]

British Columbia provides a combination of tax rebates in order to make its carbon tax revenue- neutral. The government provides a personal income tax rate cut, a low-income "climate action tax credit," a small business rate cut, a general corporate tax rate cut, and industrial and farm property tax cuts. In addition, British Columbia distributed a one-time check for C$100 to residents in June 2008.[14]

The effects of British Columbia's carbon tax are still unknown because the tax was only recently implemented. However, it was estimated that the tax would reduce emissions by 3 million metric tons annually by 2020 (Ministry of Finance, British Columbia 2008).

Bay Area Air Quality Management District, California—2008

The Bay Area Air Quality Management District (BAAQMD) incorporates nine counties of the San Francisco Bay Area. The BAAQMD established a carbon fee in July 2008, and in June 2009, it was increased by 3% to $0.045 per metric ton of carbon dioxide equivalent (CDE). The fee applies to GHG emissions from BAAQMD permitted facilities. The BAAQMD establishes the cost of implementing GHG reduction programs and then sets the rate by dividing the cost by the total amount of GHG emissions from BAAQMD permitted facilities (Bateman 2009). Approximately 780 facilities are subject to the fee (Bateman 2009). The GHG fee raises revenue for BAAQMD Climate Protection Program projects related to stationary sources. Funded activities include completing and maintaining a regional GHG emissions inventory, supporting local efforts to reduce GHG emissions from stationary sources, developing regulatory measures for GHG emissions from stationary sources, reviewing GHG- related documents, addressing climate issues in the California Environmental Quality Act, and performing administrative activities such as updating databases and invoicing (BAAQMD 2008). The fee is expected to raise $1.1 million for the BAAQMD (Bateman 2009). It may be raised in the future based on the funding needs of the BAAQMD Carbon Protection Program.

France—Proposed

In September 2009, French President Nicholas Sarkozy proposed a carbon tax of $24.74 (€17) per metric ton of CO_2 on fossil fuels such as gasoline, gas, and coal (Sarkozy 2009). Sarkozy proposed that the tax would not apply to electricity, saying that France's electricity is primarily carbon free (approximately 80% nuclear and approximately 10% hydroelectric). In addition, the electric sector is covered by the EU's ETS. The proposed price is based on the current market price for CO_2 emissions permits through the EU (Keller 2009). The tax would add $0.07 (€0.045) for each liter of diesel, $0.06 (€0.04) for each liter of gasoline, and $0.006 (€0.004) for each kWh of natural gas consumed (Keller 2009). Sarkozy (2009) expects the tax to raise $4.499 billion (€3 billion). Tax rates are expected to increase over time; however, the level of increase has not been presented. The proposal indicated that the tax would be accompanied by a reduction in taxes for households and businesses or repaid through a "green check" (Keller 2009).

California Air Resource Board—Proposed

The California Air Resources Board (CARB), under the authority of California's Assembly Bill 32 (AB 32), the Global Warming Solutions Act of 2006, adopted a carbon fee in September 2009 and is in the rulemaking process to institute the fee (CARB 2009a). The CARB staff proposal for the fee would cover 85% of California's GHG emissions through regulation of approximately 350 sources, including large natural gas distributers and large users of natural gas, producers or importers of gasoline or diesel fuel, facilities that combust coal and petroleum coke, refineries, cement manufacturers, and electricity importers and in-state generating facilities (CARB 2009b). The preliminary proposed costs for the fee in fiscal year 2010-2011 are $0.155 per metric ton CO_2 equivalent. CARB established the fee by calculating the revenue requirement to implement GHG reduction programs, and then dividing by the estimated GHG emissions by covered sectors. The fee was not designed to place a value on carbon high enough to change behavior but rather to provide a continuous funding source to implement AB 32. The fee is expected to raise $63.1 million during the first three years of implementation, in part to cover loans the programs have received in fiscal years 2007-2008, 2008-2009, and 2009-2010. The fee is proposed to be reduced to $0.09 per metric ton of CO_2 in 2014—after loans are paid off—and generate an estimated $36.2 million per year thereafter. The fee is not designed to change purchase behavior but would fund the administration, implementation, and enforcement of GHG emission reduction measures in the AB 32 scoping plan, which was designed to help California meet its GHG reduction goals (CARB 2008). In this way, the CARB fee more closely represents an administrative cost-covering fee than a traditional carbon tax. The polluting entity in this system is paying to cover the costs to administer a program to regulate emission. (See Ekins 1999)

4. Effectiveness of Carbon Taxes

While the primary purpose of carbon taxes is to reduce GHG emissions, most existing carbon policies introduce no processes or specific requirements to evaluate policy effectiveness in reducing emissions, although some attempted to assess their effects. Determining the overall impacts of a carbon tax can be challenging because many factors can affect overall carbon dioxide emissions, including economic growth levels and other programs designed to address environmental impacts.[15] Even so, measurement, tracking, and program evaluation are important to determine the impact of the policy, especially if emissions reductions are the ultimate goal.

Evaluation of Emissions Benefits of Carbon Taxes

Jurisdictions have used a variety of metrics to determine the emission benefits of carbon taxes. One of the most rudimentary metrics for measuring carbon tax effectiveness is overall reductions in GHG emissions that can be tracked using GHG emissions inventories at the national or local level. This metric is flawed in that it captures not only the carbon tax effects but also the effects of other carbon mitigation polices and exogenous variables such as the level of economic growth. While this metric lacks precision, jurisdictions can use it to evaluate whether they are meeting overall GHG reduction goals and to determine whether policies, including carbon taxes, should be modified accordingly.

Table 6. Estimated Emissions Reductions in Jurisdictions with Carbon Taxes[a]

Unless otherwise noted, decreases in emissions represent total emission reductions, not emission reductions that are due to a carbon tax.			
Jurisdiction	**Start Date**	**Change in CO_2 Emissions**	**Source**
Finland	1990	Emissions were 7% lower in 1998 than they would have been without a tax.	Prime Minister's Office, Finland (2000)
Netherlands	1990	Emissions were expected to be reduced by 1.7 to 2.7 million metric tons CO_2 annually in 2000. In covered sectors, emissions were expected to be reduced by approximately 5%.	Netherlands Ministry of Housing, Spatial Planning and the Environment (n.d.)
Norway	1991	Emissions increased by 15%—and GDP increased 70%— from 1991 to 2008.	Abboud (2008)
Sweden	1991	Emissions were reduced by about 15% from 1990 to 1996 because of the carbon tax. Emissions decreased by 9% from 1990 to 2006. Emissions decreased by more than 40% from the mid-1970s to 2008.	Johansson (2000) Ministry of the Environment, Sweden (2008)
Denmark	1992	Emissions decreased by 15% per capita from 1990 to 2005.	Prassad (2008)
United Kingdom	2001	Emissions decreased by more than 58 million metric tons CO_2 from 2001 to 2005. Emissions are expected to be reduced by 12.8 million per year (15% of commercial and public metric tons CO_2 sector energy demand) in 2010 because of the Climate Change Levy.	Cambridge Econometrics (2005) cited in Her Majesty's Treasury (2008:101)
Boulder, CO	2007	Emissions in 2007 and 2008 decreased from 2006 levels. Greatest reductions due to programs funded by the carbon tax: • Renewable energy activities (60,000 metric tons CO_2e) • Transportation (33,000 metric tons CO_2e) • Energy efficiency (6,700 metric tons CO_2e)	City of Boulder (2009b
Quebec	2007	Emissions were expected to be reduced by 11.2 million metric tons CO_2 by 2012 due to the carbon tax.	Quebec (2008)
British Columbia	2008	GHG emissions were expected to be reduced emissions by up to 3 million metric tons CO_2 annually in 2020 due to the tax.	Ministry of Finance, British Columbia (2008)

[a] BAAQMD implemented a carbon tax in 2008 and is tracking data but has not issued a report.

France and CARB each proposed but have not implemented a program.

Examining the effects of a carbon tax alone on GHG emissions would provide a more precise estimation of policy effectiveness. Many governments model the effects of a carbon tax acting alone during the implementation phase of the tax. For example, in the United Kingdom, according to a modeling study by Cambridge Econometrics (2005), the Climate Change Levy would reduce energy demand by approximately 15% (12.8 million metric tons) in the commercial and public sectors by 2010 (Her Majesty's Treasury 2008:101). However, determining the actual impact of a tax in isolation of other factors is often difficult, and most evaluations have not attempted to do so. Because of the lack of common evaluation practices, it is difficult to compare the effects of policies across jurisdictions. Table 6 summarizes evaluations of either carbon emissions levels or program effectiveness that jurisdictions have conducted to gauge the impact of carbon taxes. For example, Sweden's emissions were reduced by almost 9% between 1990 and 2006, while in Denmark, per capita emissions were reduced by 15% between 1990 and 2005.

Benefits of Programs Funded by Carbon Tax Revenues

The impacts of programs funded by specific carbon mitigation measures can be quantified to determine program effectiveness. With such program evaluations, governments administering carbon taxes can change the rate or shift funds to programs that are more effective at reducing emissions. For example, each year the City of Boulder evaluates programs funded through its carbon tax. In July 2009, the Boulder City Council approved an increase in the carbon tax level and targeted the additional revenue at expanding program offerings, including energy efficiency improvements and mass transit programs. (City of Boulder 2009a).

Estimating the impacts of individual programs funded through a tax may be easier than estimating the impacts of the tax itself, as most programs are targeted toward specific reductions that typically can be measured. For example, it is possible to calculate the cost per metric ton of CO_2e emissions over the lifetime of the project. Boulder used such a metric and found that energy efficiency programs provided the most cost effective programs. However, reductions from some programs, such as education and outreach, may be difficult to quantify.

Program Evaluation and Adjustments to Carbon Tax Levels

Linking a carbon tax level to the achievement of emissions goals emphasizes the need for accurate emissions tracking and evaluation. Ideally, the methodology for measuring emissions would be clearly articulated before the tax is implemented. Some jurisdictions, such as countries that are signatories to the United Nations Framework Convention on Climate Change, may already be completing greenhouse gas inventories.

Most carbon taxes were not designed to adjust depending on emissions levels. In British Columbia, however, the government specified that future changes in the tax rate will depend on whether GHG emissions targets are being met, the impacts other carbon policies are having, the actions other governments are taking, as well as general advice being offered by its Climate Action Team (Ministry of Finance, British Columbia 2008). Because the tax is new, this evaluation and adjustment have not yet been implemented.

5. Integration of Taxes and Other Carbon Policies

A government trying to achieve GHG reductions in its entire economy may implement a carbon tax in conjunction with complementary policies. Taxes can be an attractive option for addressing emissions that might be difficult to regulate through other mechanisms such as cap and trade or command and control regulation. For example, Norway's carbon tax covers about 68% of CO_2 emissions, which represents more than half of its total GHG emissions. By comparison, Norway's participation in the EU's Emission Trading Scheme (ETS) covers 35%-40% of its GHG emissions (Ministry of the Environment, Norway 2008). In addition to its carbon tax, Norway implemented a tax on the import and production of other GHGs including HFCs and PFCs, which have high global warming rate potentials. Emissions from these sources are expected to be cut in half because of this tax (Ministry of the Environment, Norway 2005b).

Carbon taxes have generally been limited to certain sectors of the economy. For example, carbon taxes in Sweden address the energy and transport sectors but rely on other GHG-reduction strategies to reduce overall GHG emissions. Sweden has added cross-sectoral instruments such as participating in the EU's ETS, investing in research and development and local programs, and developing policies to addresses the waste sector (Ministry of Sustainable Development, Sweden 2005b).

Like Sweden, Finland participates in the EU's ETS. Businesses are subject to both the ETS and the carbon tax, but to date Finland has allocated emissions allowances under the ETS free of charge (Parkkinen 2009). Finland has also developed building code regulations, waste management policies, and CO_2-based rates for vehicle registration taxes and annual vehicle taxes (Parkkinen 2009).

In the United Kingdom, businesses subject to the Climate Change Levy (CCL) can sign climate change agreements with the government to reduce emissions, and after achieving targets, can receive an 80% reduction in the CCL. One way businesses can meet their targets is through purchasing allowances through the EU's ETS (UK Department for Environment, Food and Rural Affairs 2005).

France has also looked to the EU's ETS to set its tax rate. President Sarkozy's proposed carbon tax would initially be set at the same level as the current market price for carbon in the ETS. The proposed tax would apply to oil, gas, and coal but not industries already participating in the ETS (BBC News 2009).

CONCLUSION AND OBSERVATIONS

Federal, state, and local governments are increasingly interested in instituting carbon taxes as a mechanism for addressing GHG emissions. While a handful of taxes were implemented in Europe in the early 1990s, few were adopted in the interim until a resurgence of interest on the part of local and state governments in recent years. Since 2007, new carbon taxes have been implemented in parts of Canada and the United States, while a number of others have been proposed.

Policy designs for carbon taxes vary according to the jurisdiction's policy goals. Some carbon taxes have been designed to encourage behavior changes and represent cost increases to fuel use as much as $30 per metric ton CO_2 or more. Other carbon taxes have been designed to raise revenues for specific carbon mitigation programs and, in some cases, are implemented at levels that are not likely to drive behavioral changes.

The use of revenues derived from carbon taxes varies considerably. Half of the taxes implemented to date return revenues to the government or entities subject to the tax to offset the burden, while the rest use revenues to fund either specific carbon mitigation programs or government budgets. Some policies apply a portion of revenues for each purpose.

Carbon taxes have been—and can be—used to address emissions from sectors that may not be easily addressed through cap and trade policies because of concerns about the point of regulation or emissions tracking. For example, non-stationary sources such as vehicle emissions can be difficult to address through a cap and trade system. Taxes have been most commonly applied to natural gas, gasoline, coal, electricity, and fuel oils.

Some carbon taxes target specific gaps and address areas not supported through other activities. As a result, carbon taxes can be integrated with other carbon policies in complementary ways.

Efforts to evaluate the effectiveness of existing carbon taxes have been limited. Some studies assessing carbon emissions levels in countries that instituted taxes in the early 1990s show overall GHG emissions reductions, with some as high as 15%. However, these studies have generally not attempted to account for the impact of other carbon mitigation policies. Most recently implemented carbon taxes emphasize evaluation and estimating impacts, but their effectiveness remains to be seen.

While it is possible to track and adjust tax levels over time to meet emissions goals, few jurisdictions have implemented such practices. British Columbia designed its new carbon tax with tracking and adjustments in mind, but the tax is so new, there has been no practical experience adjusting the tax level to meet emissions goals. Remaining to be seen are whether raising carbon tax levels entails political difficulties and how progress toward emissions goals will be evaluated; however, taxes designed with tracking and adjustments in mind hold promise as another means of achieving GHG emissions reduction goals.

Many questions still exist for future consideration.

- While carbon policy has seen greater interest at the federal level in the United States, to what degree will state and local governments pursue carbon taxes as a means for mitigating climate change?
- How would current and proposed carbon taxes in the United States interact with regional or federal cap and trade policies?
- When a carbon tax is set at a rate high enough to impact consumer behavior, how does the point of regulation affect consumer awareness? If the tax is imposed on upstream sources and consumers do not see a direct correlation between their consumption and the level of the tax they pay, they may be less likely to change

consumption behavior. Boulder's tax appears on consumers' utility bills, but carbon taxes in other jurisdictions may not be directly visible to the consumer.

- As carbon accounting and carbon registries become more advanced, will carbon taxes include life-cycle emissions of sources?

References

[1] Abboud, L. (2008). An exhausting war on emissions: Norway's efforts to contain greenhouse gases move forward–and backfire. *Wall Street Journal,* September 30, 2008. http://online.wsj.com/article/SB122272533893187737.html Accessed December 28, 2009.

[2] Bateman, B. (2009). "*Bay Area Air Quality Management District greenhouse gas fee schedule.*" Presented at the Center for Climate Change webinar. August 25, 2009.

[3] Bay Area Air Quality Management District (BAAQMD). (2008). "Fact sheet: Proposed Bay Area Air Quality Management District greenhouse gas fee schedule." February 5, 2009. http://www.baaqmd.gov/Divisions/Planning-and-Research/Planning-Programs-and-Initiatives/Climate-Protection-Program/~/mediaAccessed December 28, 2009.

[4] BBC News. (2009). "*France set to impose carbon tax.*" Sept. 10, 2009. http://news.bbc.co.uk/2/hi/8248392.stm Accessed December 28, 2009.

[5] Boulder Revised Code. 3-12-2. "*Imposition of climate action plan excise tax.*" http://www.colocode.com/boulder2/chapter3-12.htm Accessed December 28, 2009.

[6] California Air Resources Board (CARB). (2008). "*Climate change scoping plan: A framework for change.*" http://www.arb.ca.gov/cc/scopingplan/document/adopted_scoping_plan.pdf Accessed December 28, 2009.

[7] CARB. (2009a). "*AB 32 cost of implementation fee regulation* (HSC 38597)." http://www.arb.ca.gov/cc/adminfee/adminfee.htm Accessed November 4, 2009.

[8] CARB. (2009b). "*Proposed AB 32 cost of implementation fee regulation and proposed amendment to the mandatory reporting of greenhouse gas emissions regulation.*" September 25, 2009. http://www.arb.ca.gov/cc/adminfee/meetings/sept25_board_presentation.pdf Accessed October 8, 2009.

[9] City of Boulder. (2009a). "*An ordinance amending the climate action plan excise taxes to the maximum rates set forth in section 3-12-2, B.R.C. 1981, and setting forth details in relation thereto.*" Ordinance No. 7657. http://www.colocode.com/boulder2/chapter3-12.htm Accessed December 28, 2009.

[10] City of Boulder. (2009b). "*Community guide to Boulder's climate action plan.*" October 2009. http://www.bouldercolorado.gov/files/CAP Accessed December 28, 2009.

[11] Congressional Budget Office. (2008). *Policy options for reducing CO_2 emissions*. February 2008. No. 2930. Washington, DC: Congressional Budget Office. http://www.cbo.gov/doc.cfm?index=8934 Accessed December 28, 2009.

[12] Ekins, P. (1999). "European environmental taxes and charges: Recent experience, issues and trends." *Ecological Economics*, (*31*), 39-62.

[13] Ellerman, A. D. & Joskow, P. L. (2008). *The European Union's emissions trading system in perspective*. May 2008. Arlington, VA: Pew Center on Global Climate Change. http://www. pewclimate.org/ docUploads/EUAccessed December 28, 2009.

[14] Enevoldsen, M. (2005). *The theory of environmental agreements and taxes: CO_2 Policy performance in comparative perspective*. London: Edward Elgar.

[15] European Environment Agency. (2009). OECD/EEA database on instruments used for environmental policy and natural resources management. http://www2.oecd.org/ecoinst/queries/
Accessed November 3, 2009.

[16] Goulder, L. H. (1995). "Environmental taxation and the double dividend: A reader's guide." *International Tax and Public Finance*, (2), 157-183.

[17] Johansson, B. (2000). "The carbon tax in Sweden," Chapter 5. *Innovation and the environment: OECD proceedings*. http://books.google0CBIQ6AEwAw#v=onepage&q=&f=false Accessed December 28, 2009.

[18] Keller, G. (2009). France's Sarkozy urges carbon tax. *Breitbart*, September 10, 2009. http://www.breitbart.com/article.php?id= D9AKFI3G2 Accessed December 28, 2009.

[19] Her Majesty's Treasury. (2008). *Budget 2008: Stability and opportunity: Building a strong, sustainable future*. HC 388. London: The Stationery Office. http://www.hm-treasury.gov.uk/d/bud08_completereport.pdf Accessed December 28, 2009.

[20] Ministry of Finance, British Columbia. (2008). "Budget and Fiscal Plan 2008/09-2010/11." February 19, 2008. http://www.bcbudget.gov.bc.ca/2008/bfp/2008_Budget_Fiscal_Plan.pdf Accessed December 28, 2009.

[21] Ministry of Sustainable Development, Sweden. (2005a). *The Swedish report on demonstrable progress under the Kyoto Protocol.* Ds 2005:57. http://unfccc.int/resource/docs/dpr/swe1.pdf Accessed November 4, 2009.

[22] Ministry of Sustainable Development, Sweden. (2005b). *Sweden's fourth national communication on climate change under the United Nations Framework Convention on Climate Change.* Ds 2005:55. http://unfccc.int/resource/docs/natc/swenc4.pdf Accessed November 3, 2009.

[23] Ministry of Taxation, Denmark. (2008). "*VAT, payroll tax and environmental taxes.*" June 23, 2008. http://www.skm.dk/foreign/english/taxindenmark2008/6649/ Accessed December 28, 2009.

[24] Ministry of the Environment, Finland. (2008a). "Detailed information on environment-related taxes and charges in Finland." http://www.environment.fi/download.asp?contentid=85625 Accessed Oct. 7, 2009.

[25] Ministry of the Environment, Finland. (2008b). "Environmentally related energy taxation in Finland." http://www.environmentAccessed Oct. 7, 2009.

[26] Ministry of the Environment, Norway. (2005a). *Norway's fourth national communication under the Framework Convention on Climate Change: Status report as of December 2005.* T-1452 E. http://unfccc.int/resource/docs/natc/nornc4.pdf Accessed December 14, 2009.

[27] Ministry of the Environment, Norway. (2005b). *Norway's report on demonstrable progress under the Kyoto Protocol: Status report as of December 2005.* T-1453 E. http://unfccc.int/ resource/docs/dpr/nor1.pdf Accessed November 4, 2009.

[28] Ministry of the Environment, Norway. (2007). *Norwegian climate policy: Summary in English: Report No. 34 (2006-2007) to the Storting.* http://www.regjeringen.no/pages/2065909/PDFS/STM20062007003 4000EN_PDFS.pdf Accessed November 3, 2009.

[29] Ministry of the Environment, Norway. (2008). "Norwegian national allocation plan for the emissions trading system in 2008-2012." March 2008. http://www.regjeringen.no/upload/MD/Vedlegg/Planer/NAP Accessed December 11, 2009.

[30] Ministry of the Environment, Sweden, (2008). "*Towards a low carbon society.*" M2008.30. http://www.sweden.gov.se/content/1/c6/11/66/69/f7b9ac55.pdf Accessed Nov. 4, 2009.

[31] Netherlands Ministry of Housing, Spatial Planning and the Environment, Directorate-General for Environmental Protection. (n.d.). "The Netherlands' tax on energy: Questions and answers." http://www.wind-works.org/FeedLaws/Netherlands/NLEnergytax2004.pdf Accessed October 7, 2009.

[32] Nordhaus, W. (2008). *A question of balance: Weighing the options on global warming policies.* New Haven, CT: Yale University Press.

[33] Organisation for Economic Co-operation and Development (OECD). (2001). *Environmentally related taxes in OECD countries – Issues and strategies.* Cited in EUROSTAT (2003) Energy taxes in the Nordic countries – does the polluter pay? Final report. March. http://www.scb.se/statistik/MI/MI1202/2004A01/MI1202_2004A01_BR_MIFT0404.pdf Accessed December 28, 2009.

[34] Parkkinen, Timo (Ministry of the Environment, Finland). (2009). Personal communication, October 8, 2009.

[35] Prasad, M. (2008). "*Taxation as a regulatory tool: Lessons from environmental taxes in Europe.*" Tobin Project Conference "Toward a New Theory of Regulation." February 1-3, 2008. http://www.sociologyAccessed December 28, 2009.

[36] Prime Minister's Office, Finland. (2000). *Environmental and energy taxation in Finland – Preparing for the Kyoto Challenge: Summary of the working group report.* http://www.vnk.fi/julkaisut/julkaisusarja/julkaisu/fi.jsp?oid=130600 Accessed December 28, 2009.

[37] Quebec. (2008). "*Quebec and climate change: A challenge for the future. 2006-2012 Action plan.*" June 2008. http://www.mddep.gouv.qc.ca/changements/plan_action/2006-2012_en.pdf Accessed October 8, 2009.

[38] Repetto, R. (2001). "Yes, Virginia, there is a double dividend." University of Colorado at Denver. http://www.cudenver.edu/Academics/Colleges/SPAAccessed December 10, 2009.

[39] Sarkozy, N. (2009). "France's strategy in the fight against climate change: Speech by Nicolas Sarkozy, President of the Republic - Excerpts." September 10, 2009. https://pastel.diplomatie.gouv.fr/editorial/actual/ael2/bulletin.gb.asp?liste=20090914.gb.html Accessed November 4, 2009.

[40] Speck, S., Andersen, M. K., Nielsen, H. Ø., Ryelund, A. & Smith, C. (2006). *The use of economic instruments in Nordic and Baltic environmental policy 2001-2005.* 2006:525. Copenhagen: Nordic Council of Ministers. http://www.norden.org/da/publikationer/publikationer/2006-525/ Accessed November 3, 2009.

[41] Swedish Tax Agency (2008a). Taxes in Sweden: An English summary of tax statistical yearbook of Sweden. http://www.skatteverket.se/download/18.69ef368911e1304a625800017687/10409.pdf Accessed November 3, 2009.

[42] Torys, LLP. (2007). "Quebec government to implement carbon tax." *Torys Climate Change Bulletin.* June 12, 2007. http://www.torys.com/Publications/Documents/Publication PDFs/ CCB2007-6.pdf Accessed October 8, 2009.

[43] Turner, C. (2008). "The carbon cleansers." *Canadian Geographic.* October 2008. http://www.canadiangeographic.ca/magazine/oct08/feature_norway_carbon_tax.asp

[44] UK Department for Environment, Food and Rural Affairs. (2005). "The United Kingdom's report on demonstrable progress under the Kyoto Protocol." http://unfccc.int/resource/docs/dpr/ uk1.pdf Accessed Nov. 4, 2009.

[45] UK Department of Energy and Climate Change. (2009). "Climate change agreements: The Climate Change Levy." http://www.decc.gov.uk/en/content/cms/what_we_do/change_energyAccessed November 3, 2009.

[46] United Nations Framework Convention on Climate Change (UNFCCC) Compliance Committee. (2006). *Report of the centralized in-depth review of the fourth national communication of Norway.* CC/ERT/2006/8. http://unfccc.int/files/kyoto_protocol/complianceAccessed November 3, 2009.

BIBILIOGRAPHY

[1] Swedish Environmental Agency and Swedish Energy Agency (2007). "Economic instruments in environmental policy." February 2007. Report 5678. Stockholm: Swedish Environmental Agency. http://www.energimyndigheten.se/Global/Engelska/News/620-5678-6_webb.pdf Accessed November 3, 2009.

[2] Swedish Tax Agency (2008b). *Facts about Swedish excise duties 2008.* http://www.skatteverket.se/download/18.225c96e811ae46c823f80007801/493b14.pdf Accessed November 3, 2009.

End Notes

[1] Representative Stark (D-CA) introduced H.R.594 "Save Our Climate Act of 2009" in January 2009, Representative Larson introduced H.R.1337 "America's Energy Security Trust Fund Act of 2009" in March 2009, and Representative Inglis (R-SC) introduced H.R.2380 "Raise Wages, Cut Carbon Act of 2009" in May 2009.

[2] In this paper, unless otherwise noted, foreign currencies have been converted to U.S. dollars using contemporary exchange rates for the given year of revenue or expenditure. If no year of the tax or expenditure is stated, a current exchange rate is used.

[3] In 2009, a two-earner family of four earning C$60,000 per year would pay carbon taxes of C$60 for gasoline and C$53 for natural gas heat and hot water, but it would receive a personal income tax reduction of C$118, thus receiving a net gain of C$5. These figures assume 23 miles per gallon fuel efficiency and approximately 12,400 miles of driving per year. A senior couple earning C$30,000 per year would pay carbon taxes of C$25 for gasoline and C$70 for an oil furnace using 2,000 liters of heating oil per year. The senior couple would receive C$205 in personal income tax reduction, thus making the net gain for the couple C$110. These figures assume 19.6 miles per gallon fuel efficiency and approximately 4,300 miles of driving per year. (Ministry of Finance, British Columbia 2008)

[4] Actual numbers: 25.810 billion SEK in 2005; 24.743 billion SEK in 2006; and 25.088 billion SEK in 2007.

[5] Pelle Magdalinski (Swedish Environmental Protection Agency), personal communication, December 9, 2009.

[6] National Insurance Contributions are mandatory contributions paid by employees and employers on earnings, and by employers on certain benefits-in-kind provided to employees.

[7] Nick Hatherall (United Kingdom Climate Change and Energy Branch), personal communication, September 18, 2009.

[8] Sarah Van Pelt (City of Boulder), personal communication, October 9, 2009.

[9] Boulder performs GHG accounting in metric tons of carbon dioxide equivalent (CO_2e)

[10] Sarah Van Pelt (City of Boulder), personal communication, October 9, 2009.

[11] Robert Noel de Tilly (Senior Policy Advisor), Quebec Government, personal communication, October 8, 2009.

[12] Robert Noel de Tilly (Senior Policy Advisor), Quebec Government, personal communication, October 8, 2009.

[13] Anne Foy (British Columbia Ministry of Finance), personal communication, September 17, 2009.

[14] Anne Foy (British Columbia Ministry of Finance), personal communication, September 17, 2009.

[15] Evaluation may be easier to implement for taxes designed to raise revenues to implement specific carbon reduction programs.

INDEX

A

B

C

D

E

F

G

H

I

N

O

P

R

S

T

U

V

W

Y